ABACUS MASTER

Tests

Abacus an Introduction

As far as modern scientists can tell; we humans first counted on our fingers and toes. Early humans had no concept of infinity or an empty set. As trading items of equal value but in different volumes, such as 2 pelts for one arrow, it became necessary to keep track of quantities and values of inventory. A trader needed to know that he or she was getting a fair and equitable trade. Researchers tell us that Man needed to record numbers greater than 20 fingers and toes and may have laid out sticks and stones as a temporary record. They also say that lines made in a dust or sand table were used. The Greek word for soil is pronounced "*edafos*", and another word "abax", counting table, were precursors to the word we use now "abacus".

The oldest abacus known is Sumerian, made more than 2,300 years ago. This device was once the primary accounting tool for many forms of business.

Several cultures developed their unique style of abacus. Cultures in Asia, India and Middle East still use the abacus to develop mental mathematical skills. Even some merchants continue to rely on the abacus.

One version of a Russian abacus has a rectangular frame with several rods running horizontally and beads of two or more colors strung on the rods.

Счеты

(Schety)

A Japanese abacus may have a rectangular frame with a horizontal bar, with one bead above and four beads below the bar, on each of several vertical rods.

そろばん

(Soro ban)

We will use the Chinese Abacus having a rectangular frame and thirteen, more or less, vertical rods or wires, and a horizontal bar dividing the frame into two decks. Each rod, or wire, holds two beads in the top deck and five beads in the bottom deck.

算盘

(Suànpán)

Each student should work the problems and examples in this book, with a Chinese Abacus.

Some abaci are antiques and many people have extensive collections of them. Some public libraries may have an abacus foe lending. An actual three dimensional abacus is better than a virtual abacus found on the internet.

It is not a bad idea to learn, and teach, how to operate more than one style of abacus but here we will concentrate on the Suanpan.

Here is a basic Chinese Abacus.

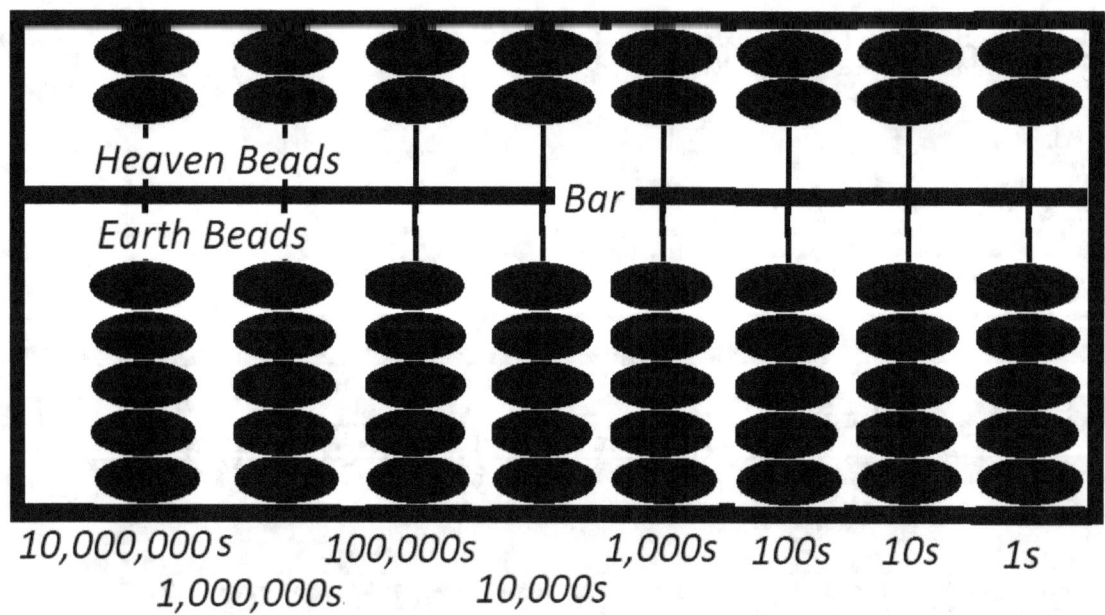

These are the place values for each column

An abacus may have any number of columns, colors, or bead values, as long as there is a consistent relationship between beads, columns and place values.

A Bead in the top deck is worth 5 times its place value. A bead in the bottom deck is worth 1 times its place value.

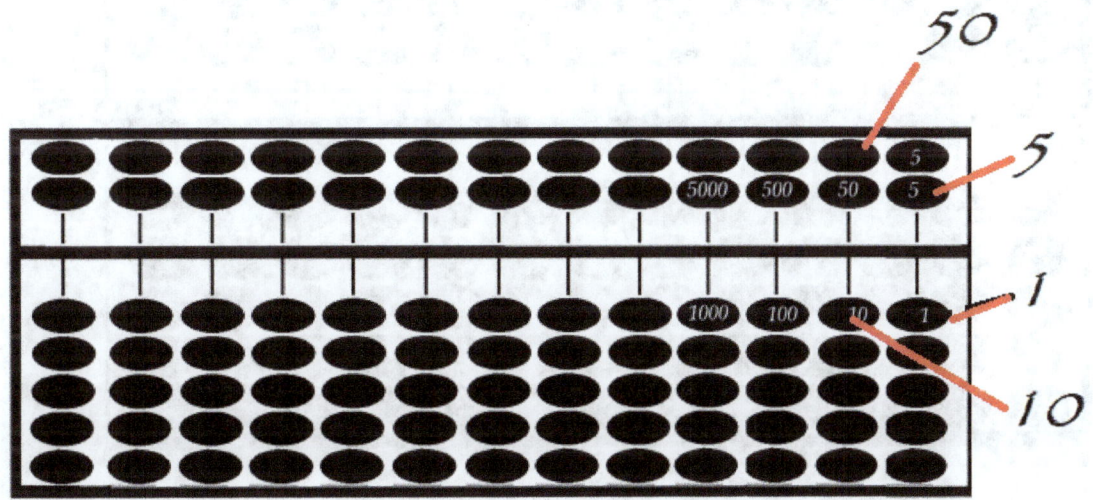

The beads in the top deck are called "Heaven Beads" and those in the bottom are called "Earth Beads".

A Heaven Bead in a column is worth five Earth Beads in the same column.

An Earth Bead in a column to the left is worth two Heaven Beads in the next column to the right.

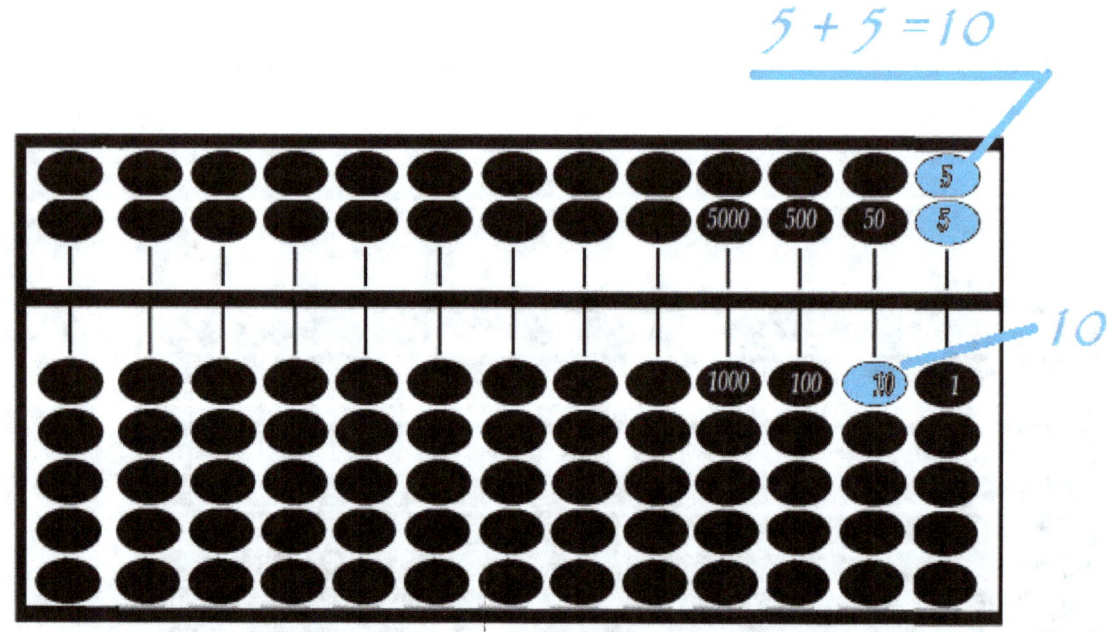

To place all the beads in a starting position, we lay the frame flat, bottom deck toward us, and then tilt the frame toward us. This brings all the Earth Beads to the bottom, and all the Heaven Beads to the bar. We run a finger along the top of the bar to push all Heaven Beads to the top of the frame, away from the bar.

With no beads touching the bar, the value displayed is zero, or starting position.

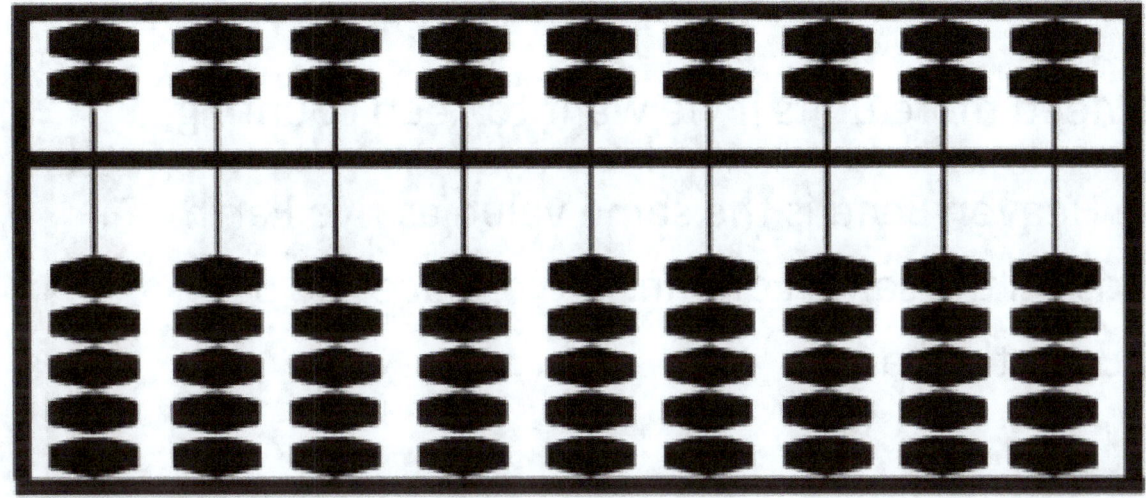

Beads that are against the bar represent numbers of 1s and 5s Look at the rightmost column. Move the Earth Beads up toward the bar as you count them.

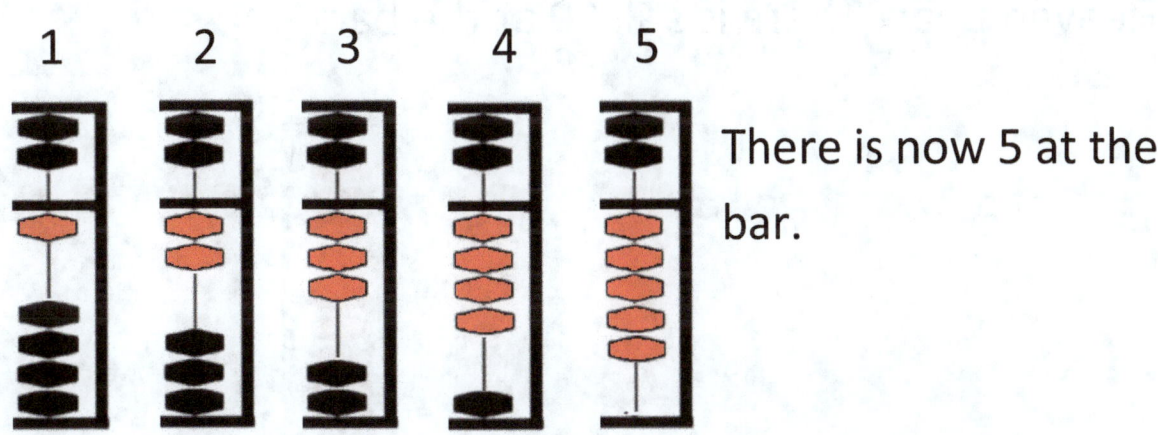

There is now 5 at the bar.

We need more units if we want to keep counting.

One Heaven Bead is the same value as five Earth Beads on the same column. We replace the 5 beads at the bar.

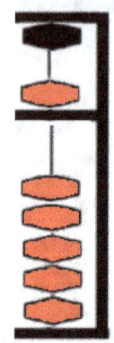

Continue the count.

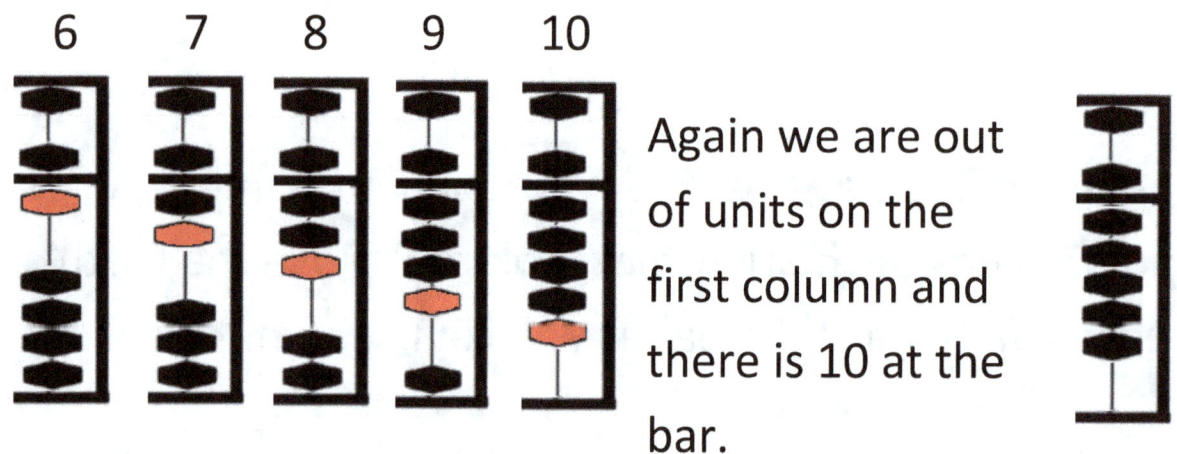

6   7   8   9   10

Again we are out of units on the first column and there is 10 at the bar.

The 5 Earth Beads are again replaced by another Heaven Bead. There is still 10 at the bar.

Count the Earth Beads a third time, to 15.

Now with 15 at the bar and no Heaven Beads available on the column, we may remember that one Earth Bead on the next column left is worth Two Heaven Beads on the next column right.

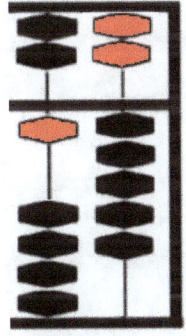

The two Heaven Beads are replaced by the one Earth Bead. There is still 15 at the bar.

Now there are Heaven Beads available to replace Earth Beads on the same column.

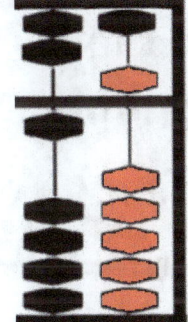

There is still 15 at the bar.

Every time we create a column with two Heaven Beads, we must replace them with an Earth Bead from the next column left.

Every time we use all five Earth Beads on a column, we must replace them with a Heaven Bead from the same column.

Make the necessary replacements:

After adjustments you should read 10,410

All the beads on a Chinese abacus are usually the same size, shape and color. I will use multi-colored beads in some diagrams and vectors to show bead movement more clearly. Remember; only the beads against the bar are counted. Special attention should be given to those columns holding a 0 value as they may be confused with empty or unused columns.

Addition

When we add we simply move beads toward the bar.

Set 13 in columns L and M.

(Red shows which beads are moved.)

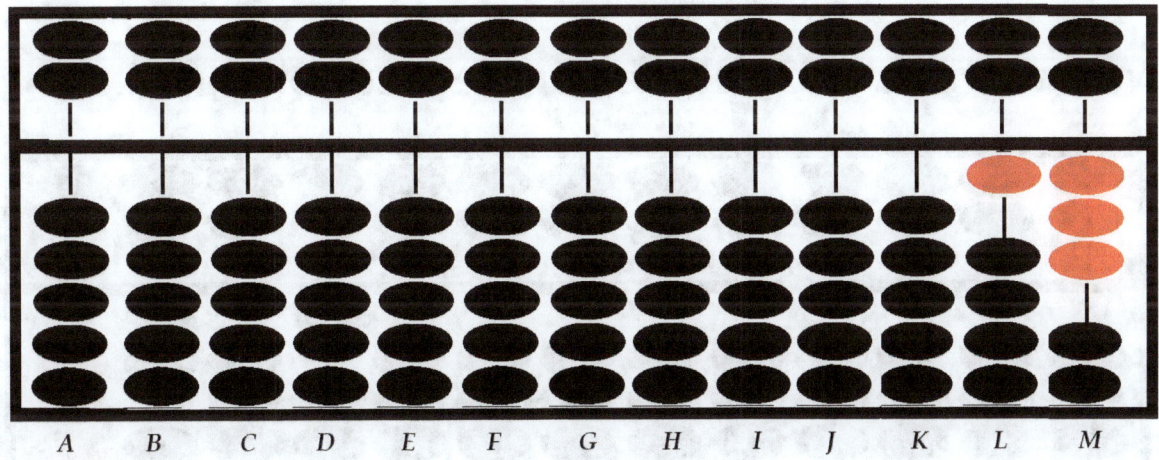

We can add to 13 by moving additional beads to the bar. Add 12 to the 13 at the bar.

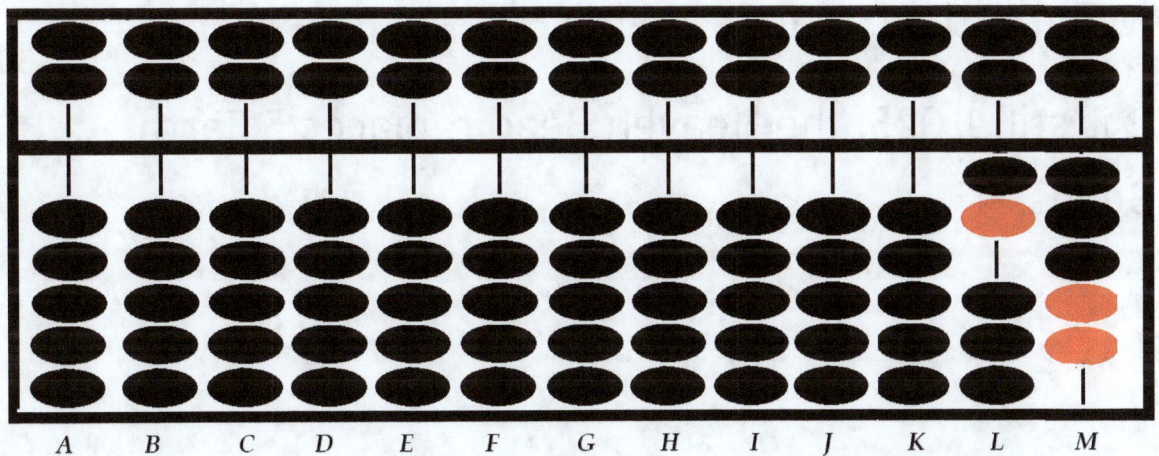

The number at the bar is now 25. Let's add 1000 to 25.

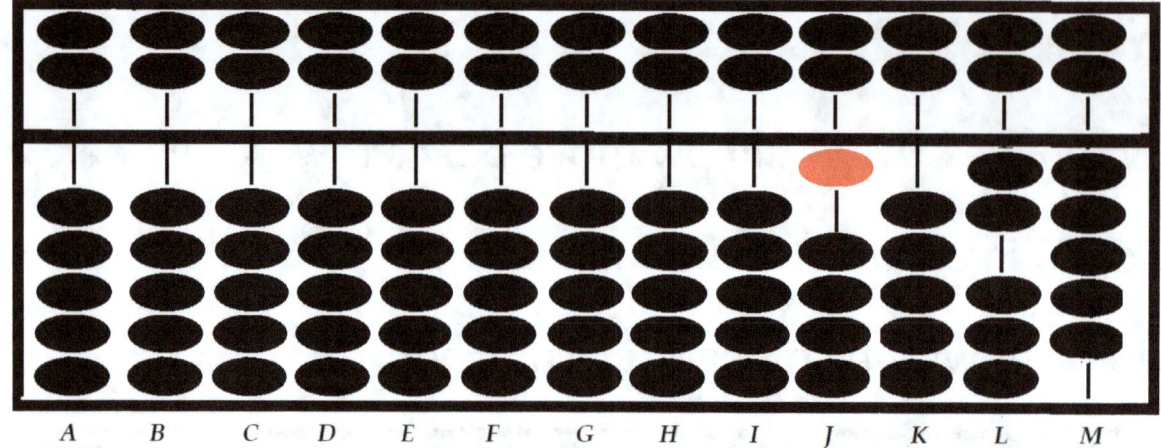

Now we have 1,025 at the bar. Reduce column M.

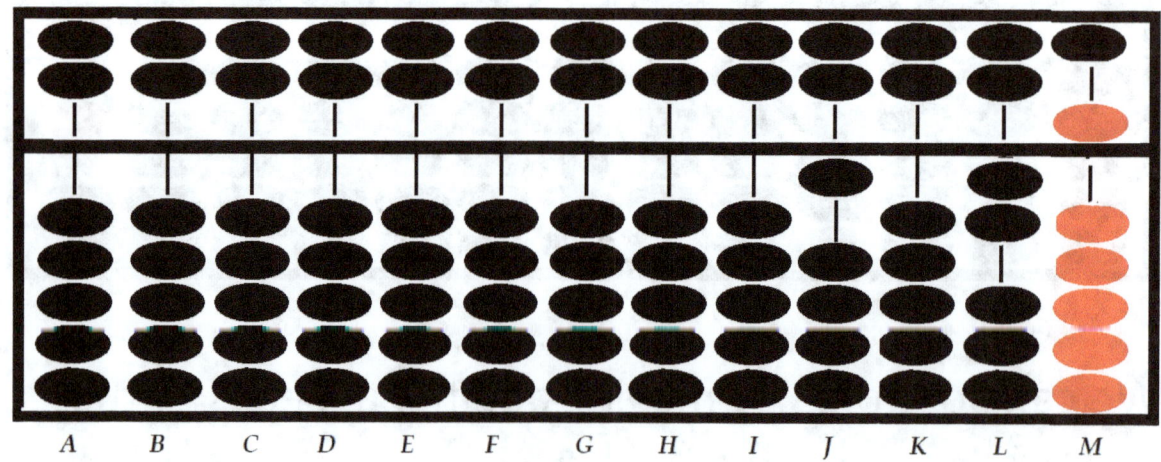

 What is the count at the bar?

It is still 1,025; the Heaven Bead replaces 5 Earth
Beads.

Add 35 to what we have now.

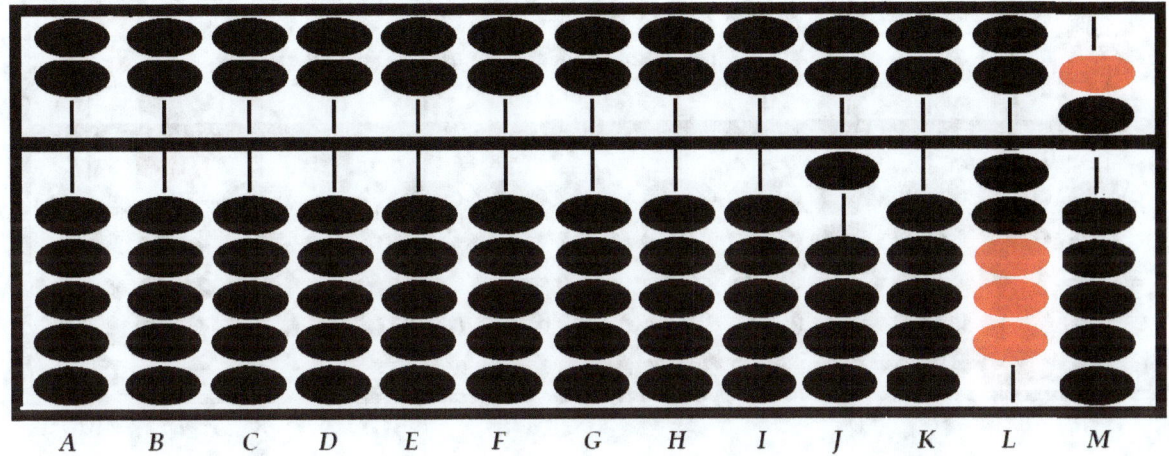

At the bar is 1,060. Reduce column L.

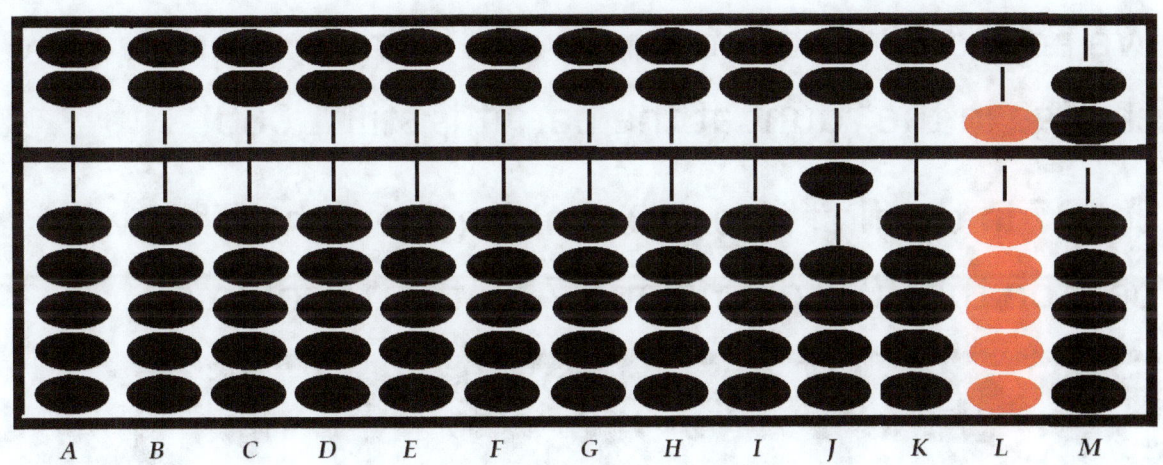

Reduce column M.

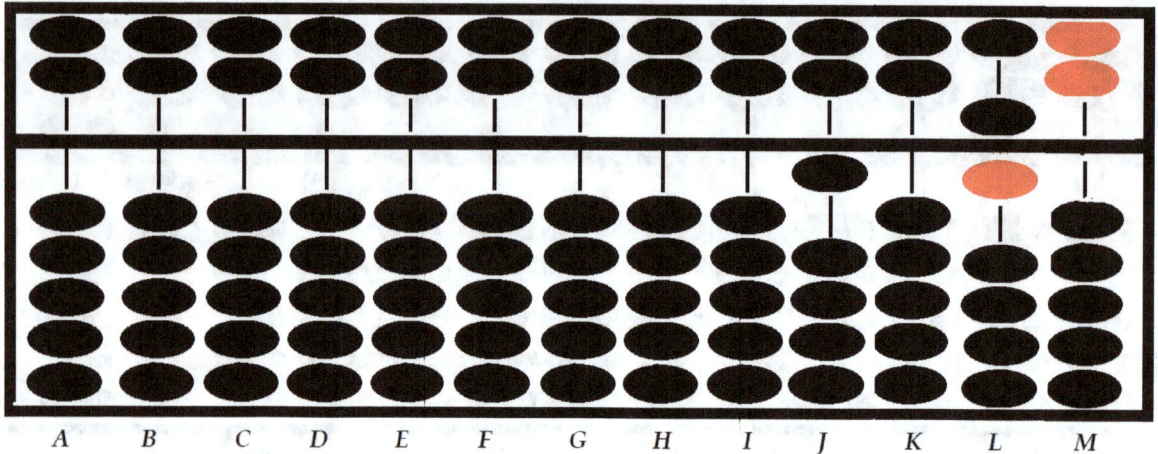

The Earth Bead on L is a replacement for the two Heaven Beads on M, which are moved away from the bar.

We added 10 on L and removed 10 on M; there was no change in the count at the bar; it is still 1,060!

Set 27 in the first two columns on the right. That will be 2 on the 10s column and 7 on the 1s column.

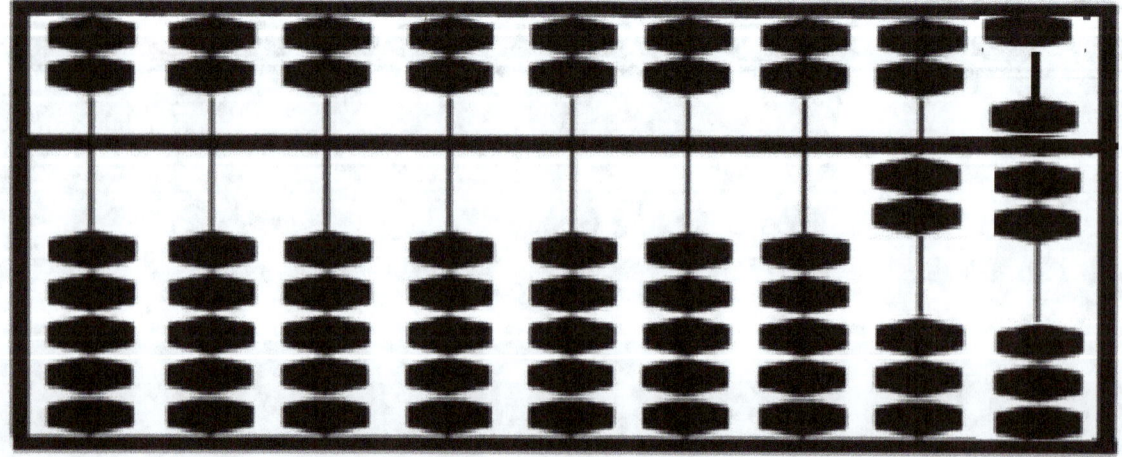

Now we will add 49 to what we have.

We need 1 Heaven Bead and 4 Earth Beads to make 9. We move an Earth Bead up on the 10s, and we move an Earth Bead down on the 1s. (+10-1=+9).

We do not have 4 beads to move up for 40 on the 10s column, so we move down 1 Heaven Bead and move down 1 Earth Bead on the 10s, (+50 -10= +40).

We can read the beads at the bar as, 7 and 6.

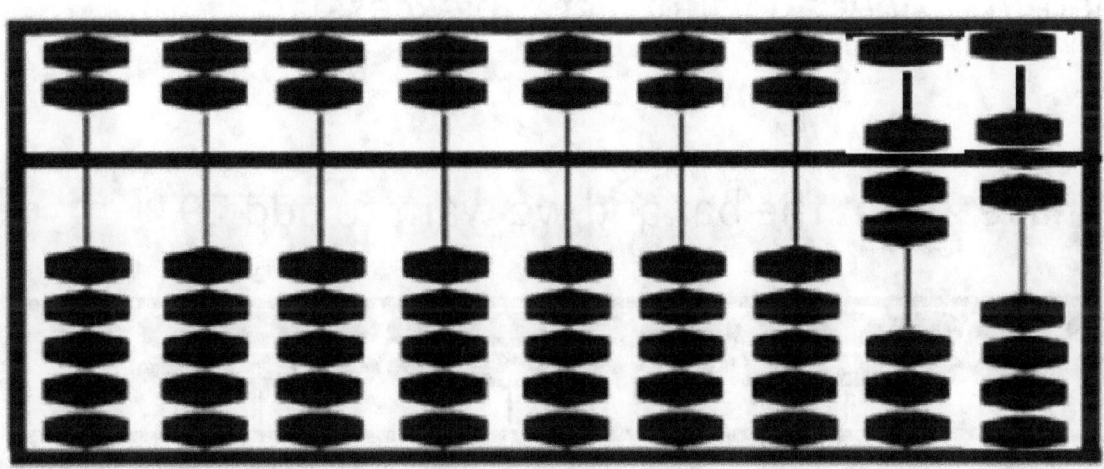

(27 + 49 = 76)

Because we post additions as we go; we don't carry values, we place them on appropriate columns.

When adding 27 to 49 the 9+7 =16 so, we post 6 to the 1s column and 1 to the 10s column. Then when we add the 4+7=11, we post another 1 in the 10s and the 1s columns.

For Addition, we simply add the beads needed for the number being added to the beads of the current number.

If we need to add units that are not available on the column we need to place them on, then we add a greater number and subtract the excess!

We have 248 at the bar and we want to add 29.

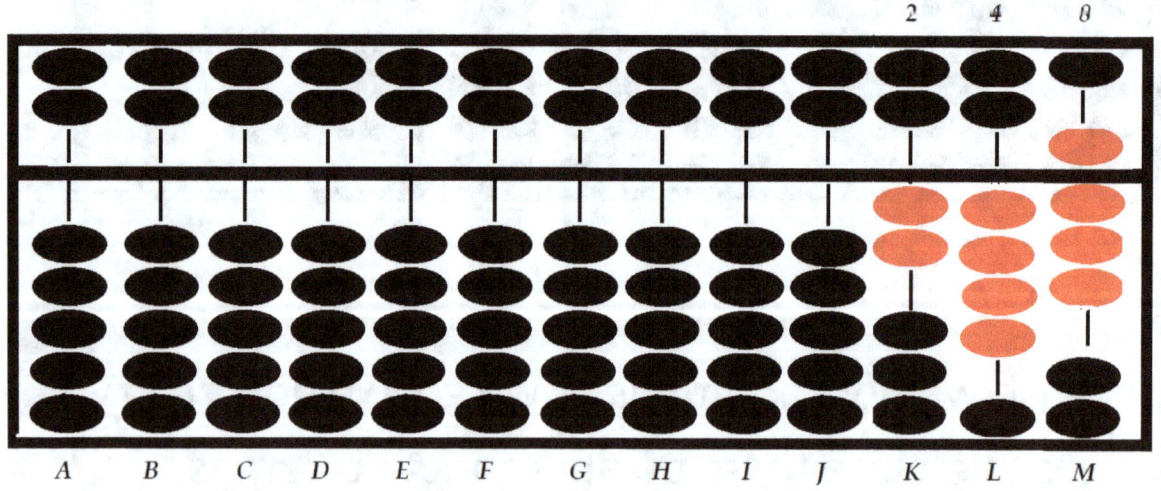

M has only 2 units available so, to add the 9, of 29, to M, we add 10 at L and remove 1 at M.

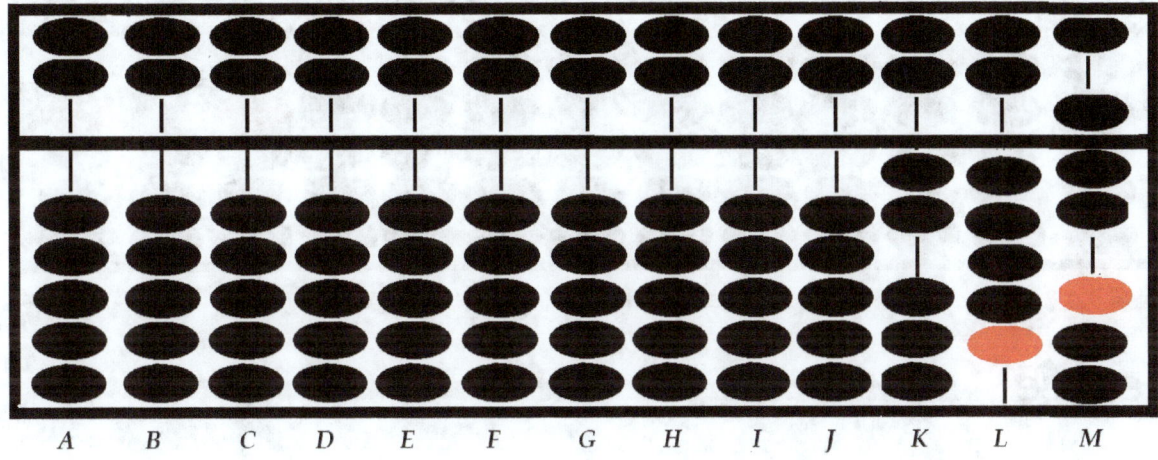

(+10 -1 = + 9) Then we need to add 2 to column L.

L has no Earth Beads away from the bar, so we must first reduce L then add the 2 to L, or a faster way is to add 5 to L and remove 3 from L.

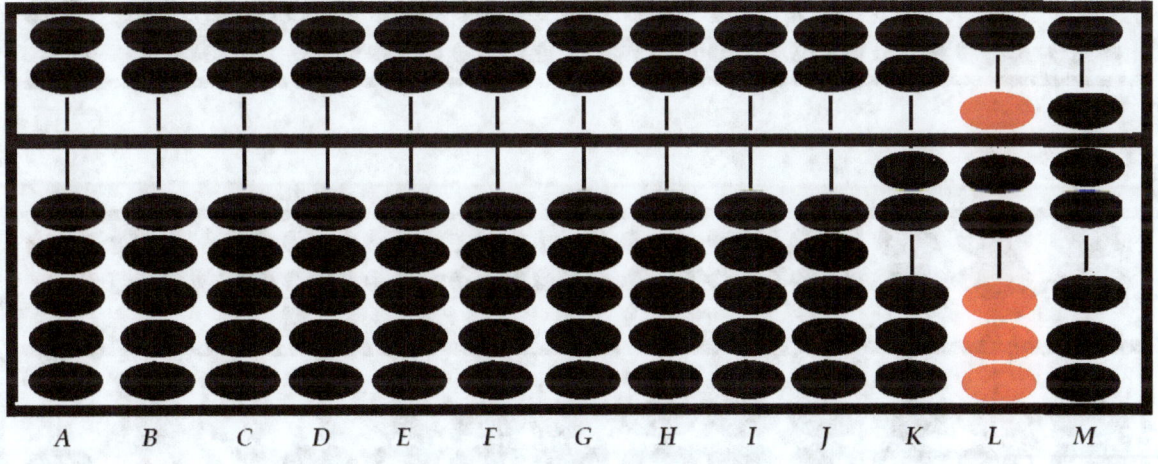

The beads now read 277, (248 + 29).

With 277 on your abacus, add 954.

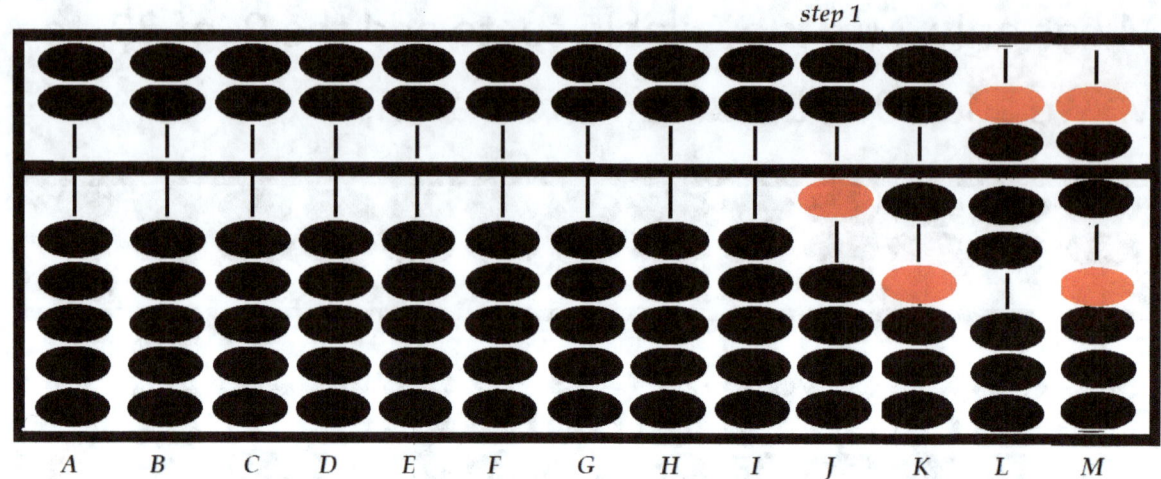

Reduce

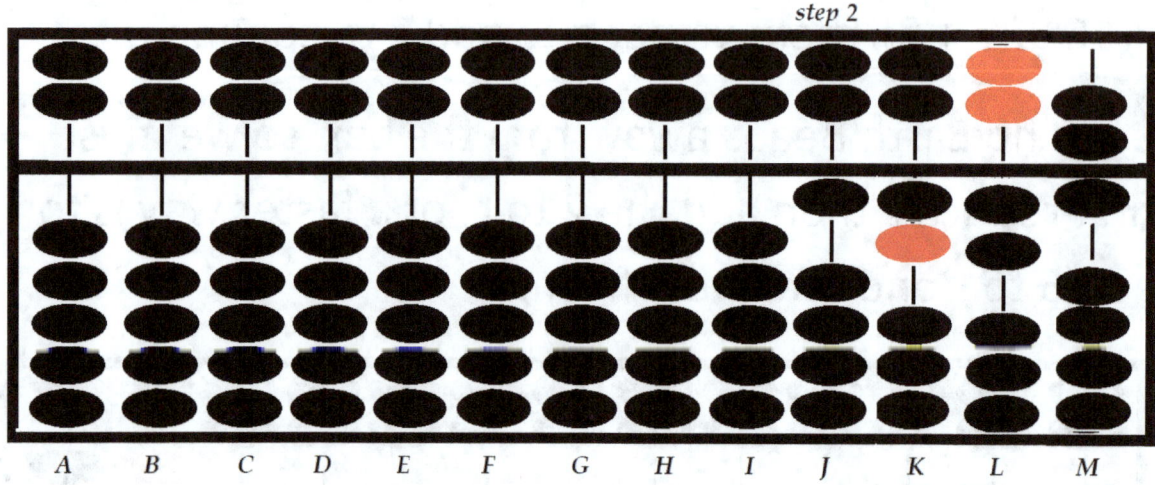

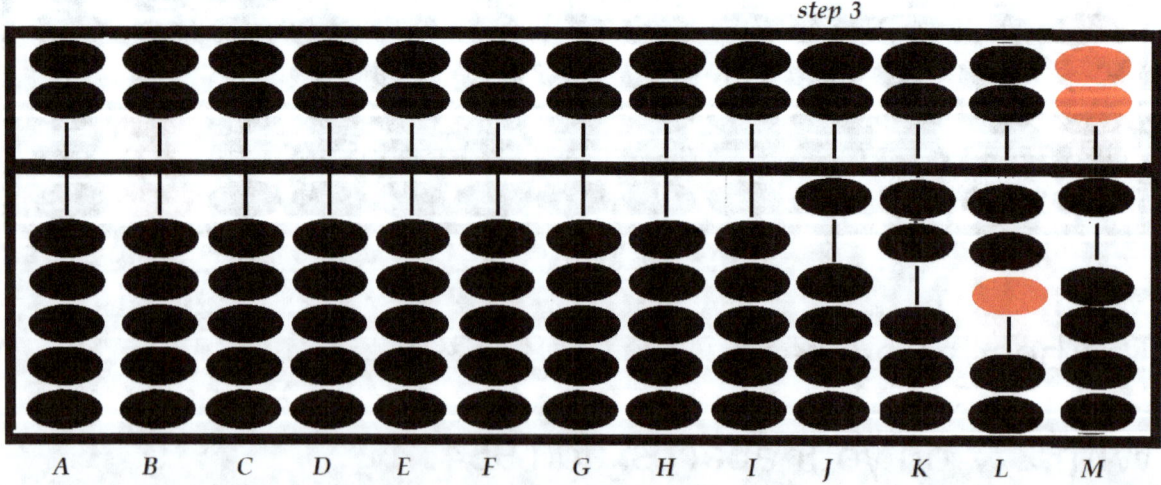

The sum is 1,231 and it should look like this.

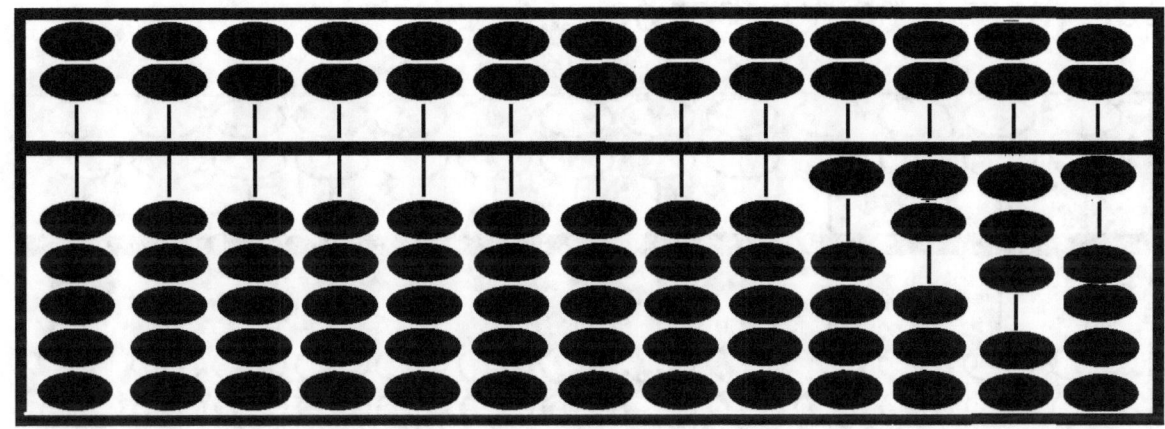

Try these on the abacus:

6+8=   12+7=   18+9=   20+14=

175+22=   321+99=   500+108=

Counting & Addition Review

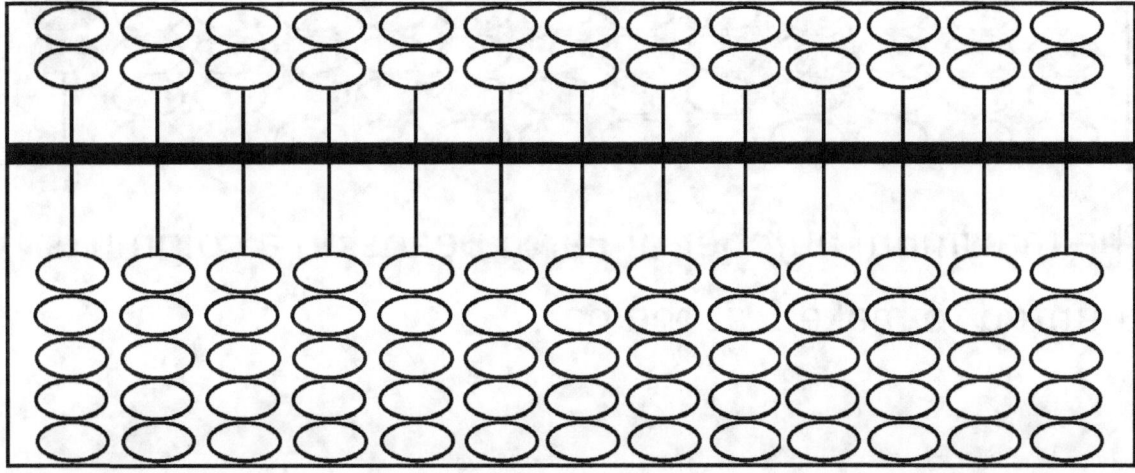

A number is represented by the beads touching the bar, there is no number here.

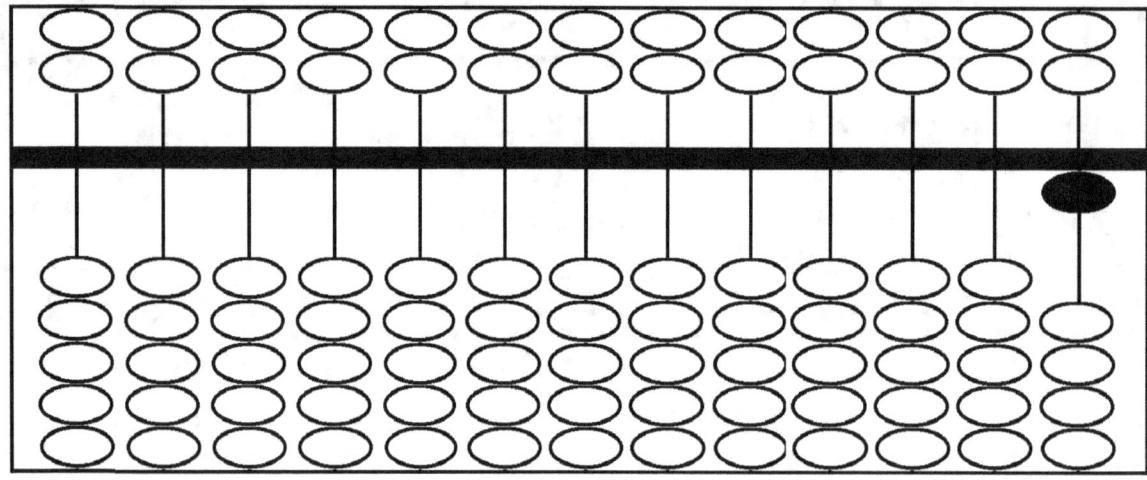

We have moved a bead to represent 1 on the rightmost column.

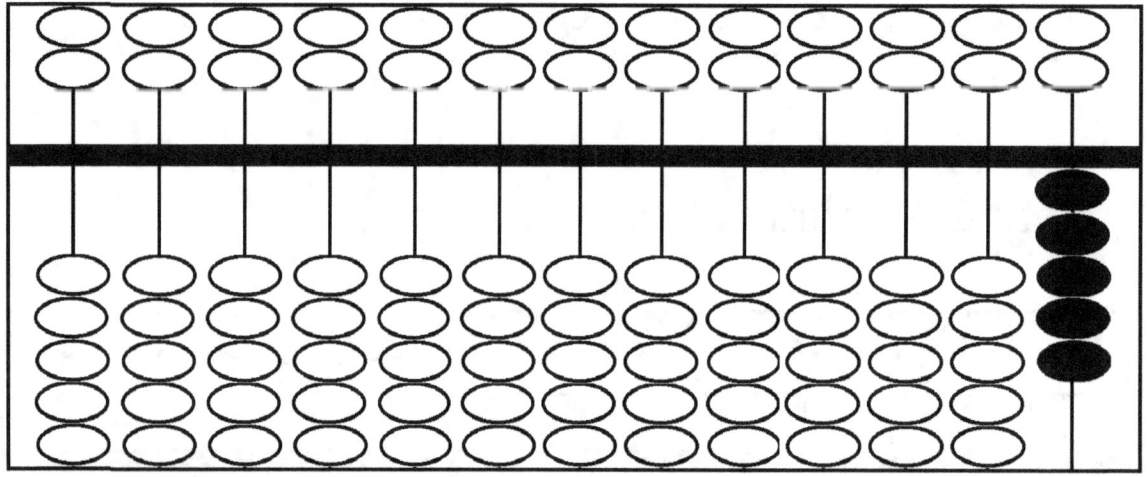

The maximum number of Earth Beads on a column is counted to make 5 at the bar.

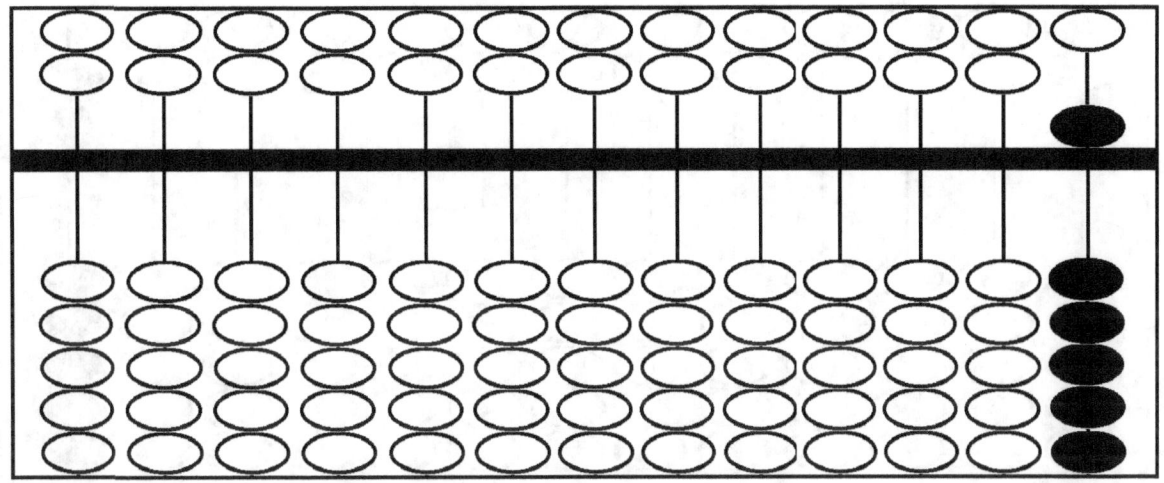

Each Heaven Bead is equal to 5 Earth Beads on the same column, so we can replace the Earth Beads with a Heaven Bead and continue counting with the Earth Beads.

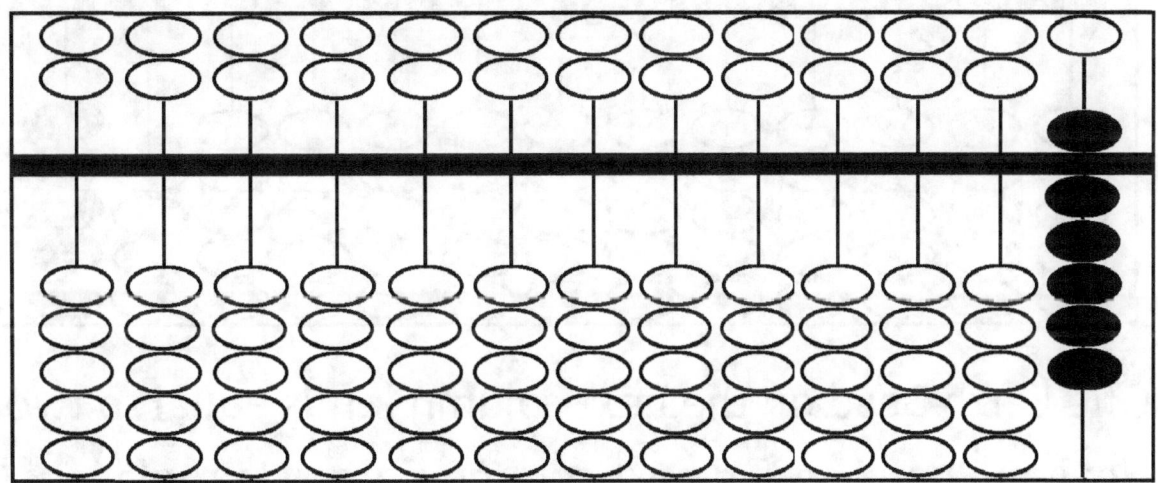

Again, having reached 10 at the bar, we replace the Earth Beads.

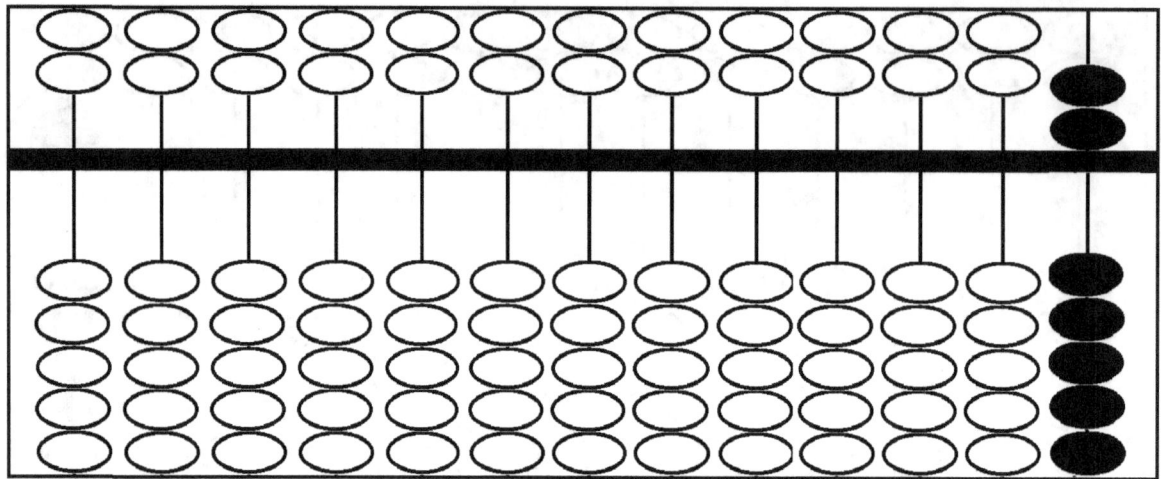

We may use the same Earth Beads to continue a third time. The count is 10 at the bar.

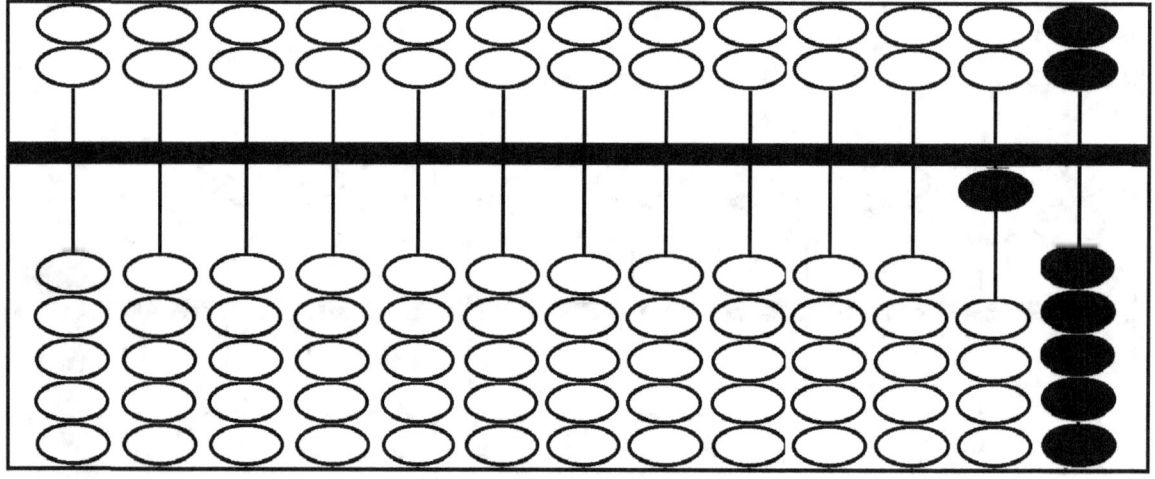

One Earth Bead on the next column left is equal to two Heaven Beads on the next column right. We replace the two Heaven Beads with one bead on the next column. The count is still 10 at the bar.

Try these:

439 + 197 =   90 + 139 =   208 + 780 =   605 + 82 =

277+1803 =   308 + 140 =   2010 + 3030 =

935 + 333 =   910 + 106 =   1650 +31416 =

108 + 1731 =   1040 +1099 =

Read these:

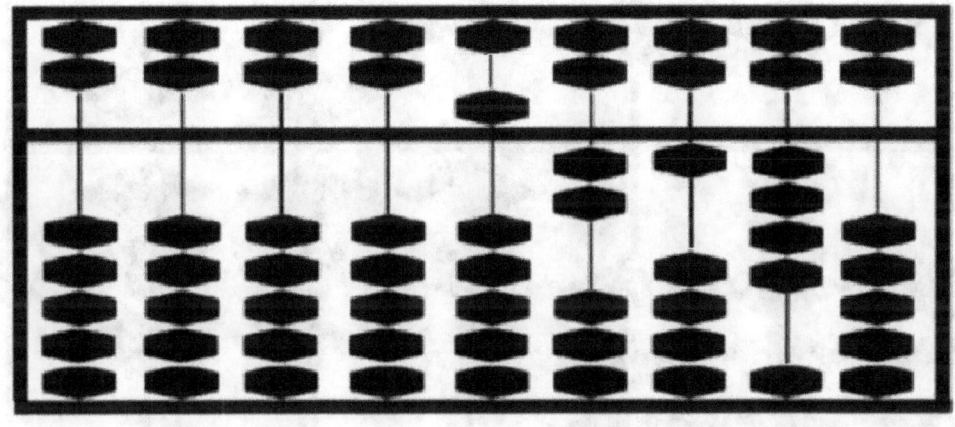

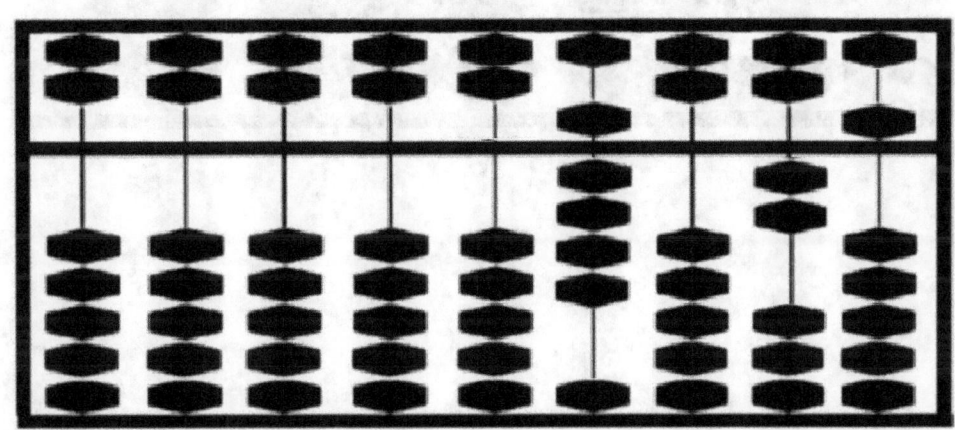

Subtraction

To subtract we move beads away from the bar.

Sometimes when subtracting a number we need to do the opposite replacement we do when adding. We may need to subtract a larger number than we intend and add back the excess!

Set 388 on the abacus

A   B   C   D   E   F

Subtract 159

First we can move an Earth Bead down on D (-100), then on column E we can move a Heaven Bead up

(-50), but to subtract the 9, we move an Earth Bead down on column E (-10) and move an Earth bead up on column F (+1).

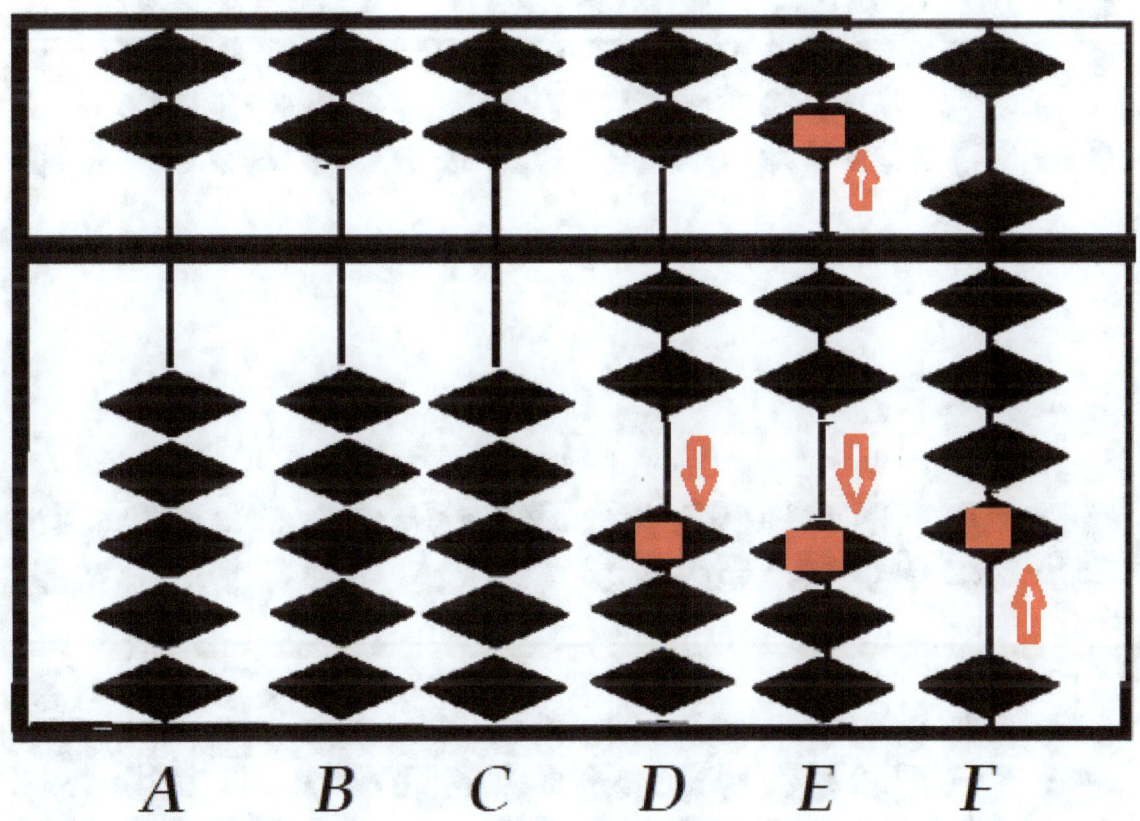

The result is 388-159 = 229.

Subtract 14,198 from 30,264

Note: no decimal places

30,264 - 14,198   set the minuend

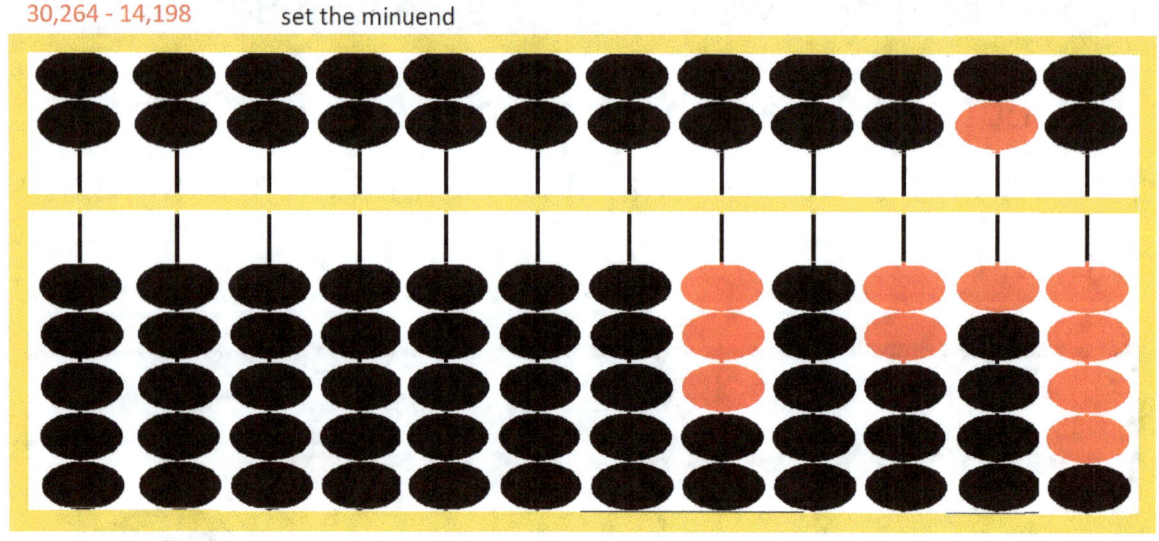

30,264 - 14,198   start on the right and work to the left

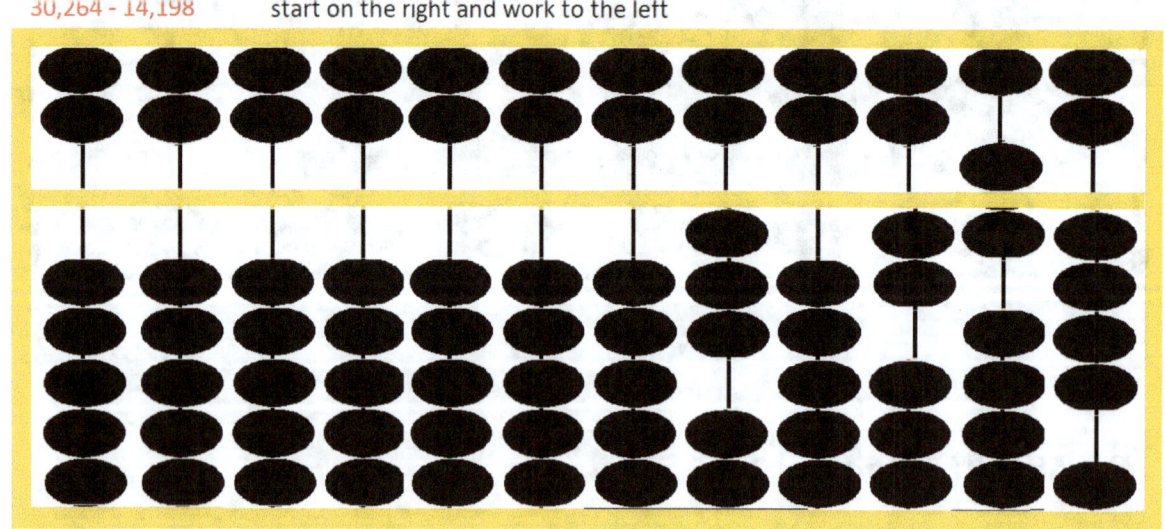

8 from 4 is not possible so we barrow from the next
column left and beads above.

30,264 - 14,198   - 10 + 5 - 3 = -8

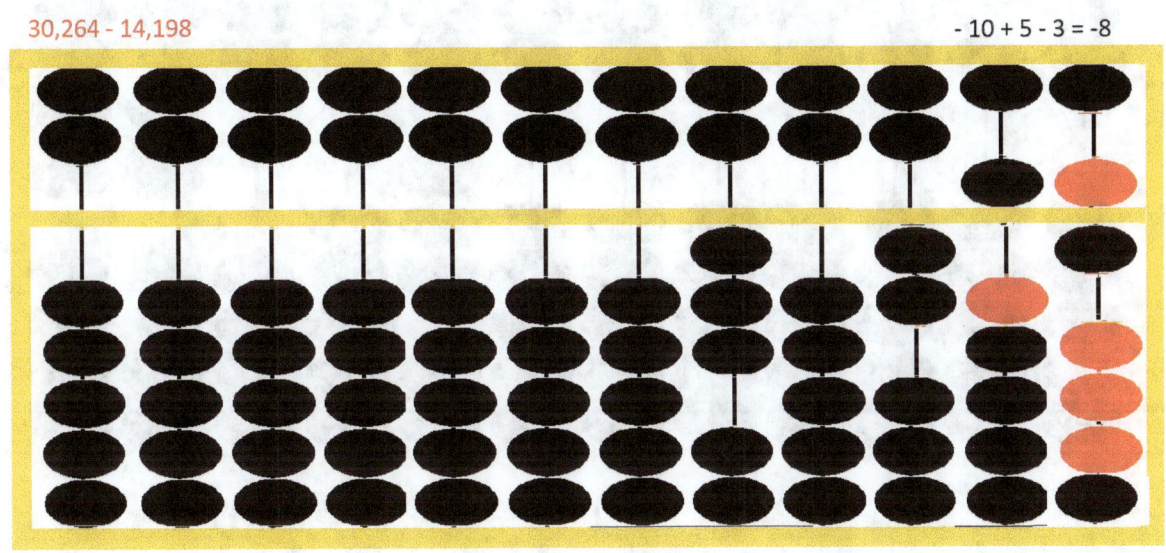

9 from 5 can't be done, so we barrow again.

30,264 - 14,198   - 100 + 10 = -90

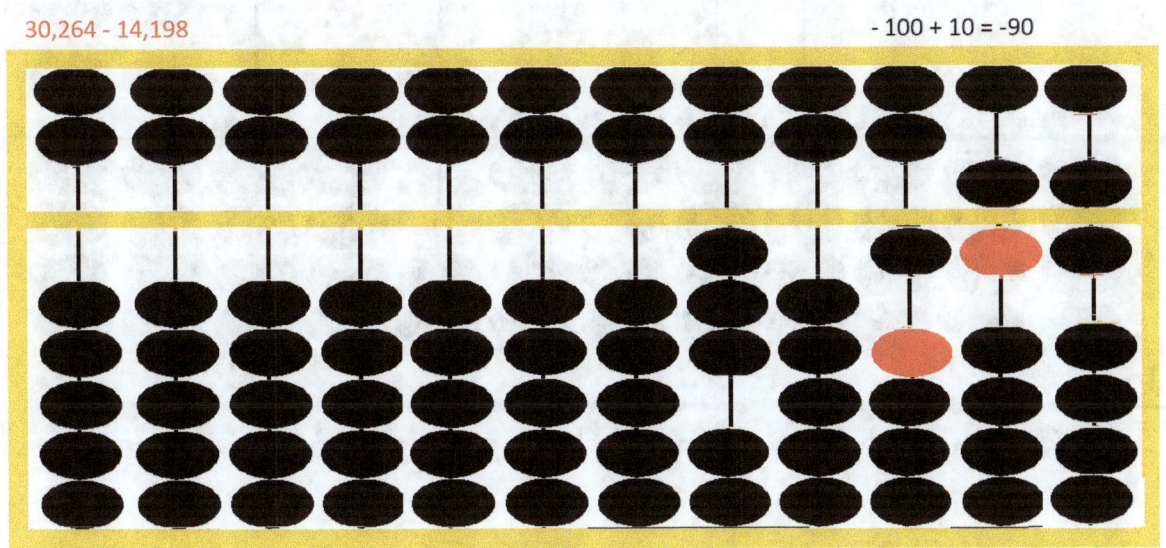

Next we subtract 100

30,264 - 14,198   - 100

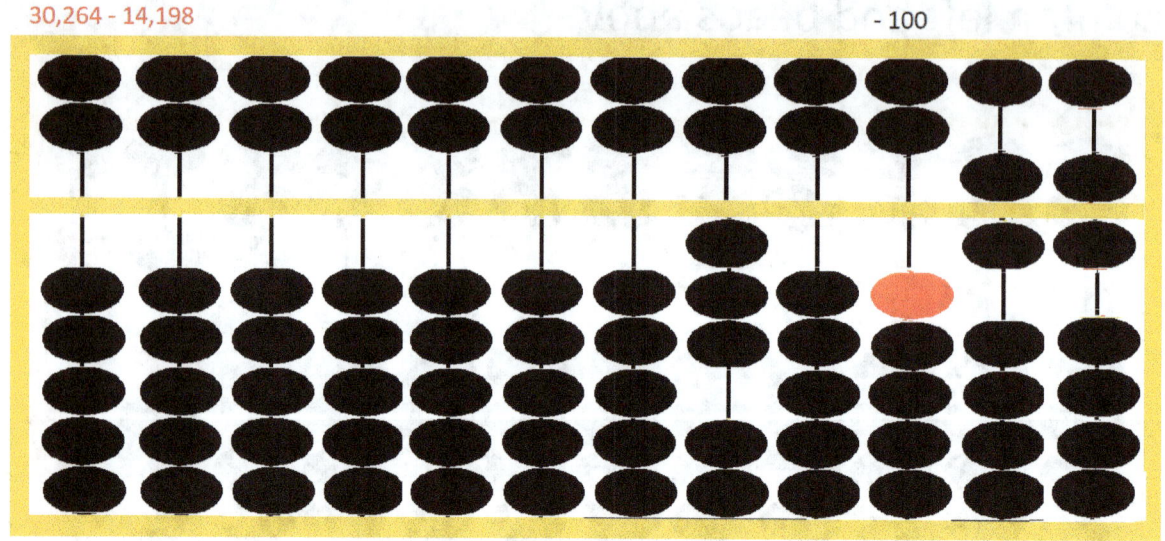

We can't take 4 from 0, so we barrow another 10

30,264 - 14,198   -10,000 + 6,000 = - 4000

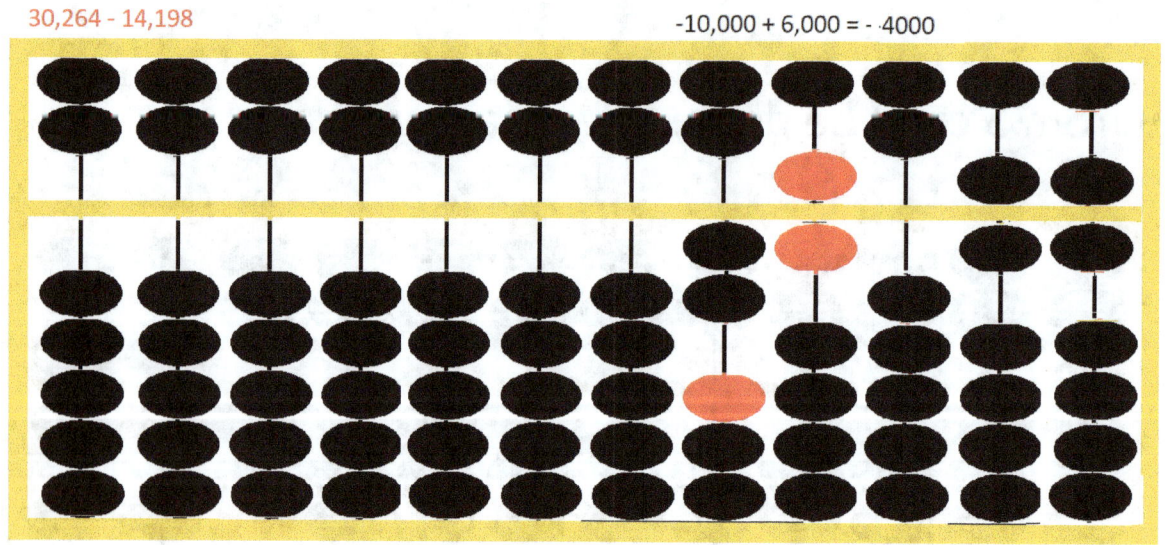

And finally the last digit of the subtrahend

- 10, 000

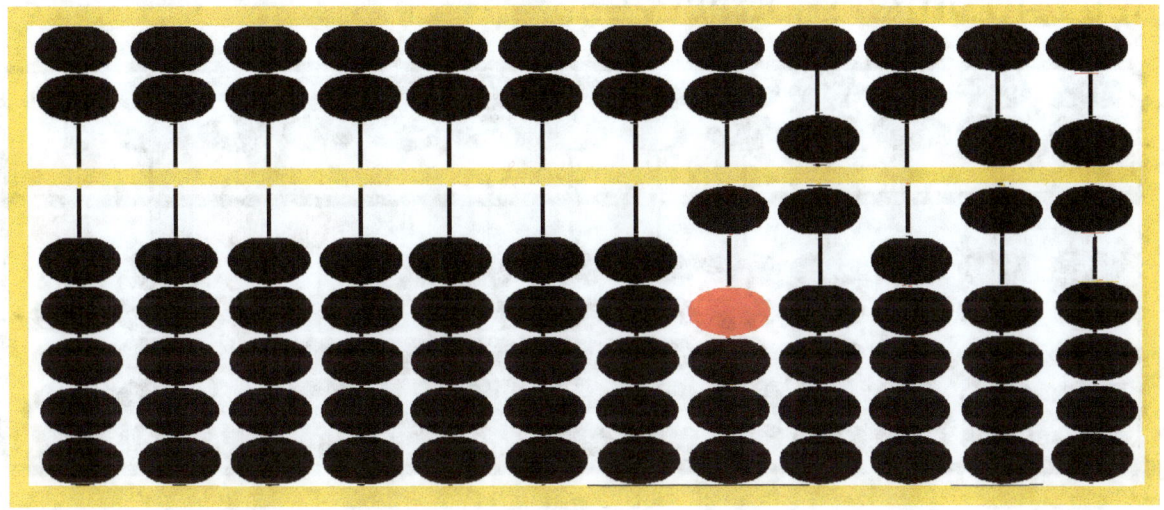

30,264 − 14,198 = 16,066

30,264 - 14,198   = 16, 066

Set 1,406 then subtract 795.

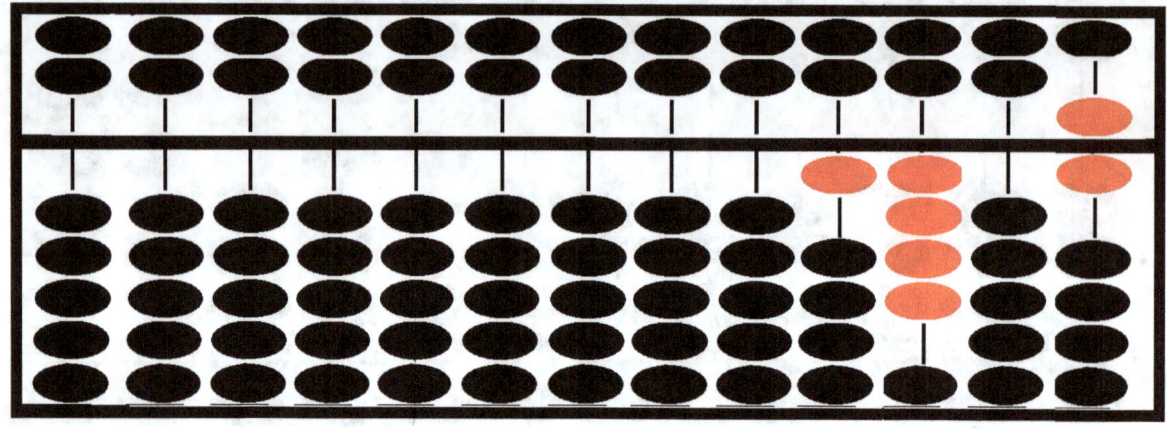

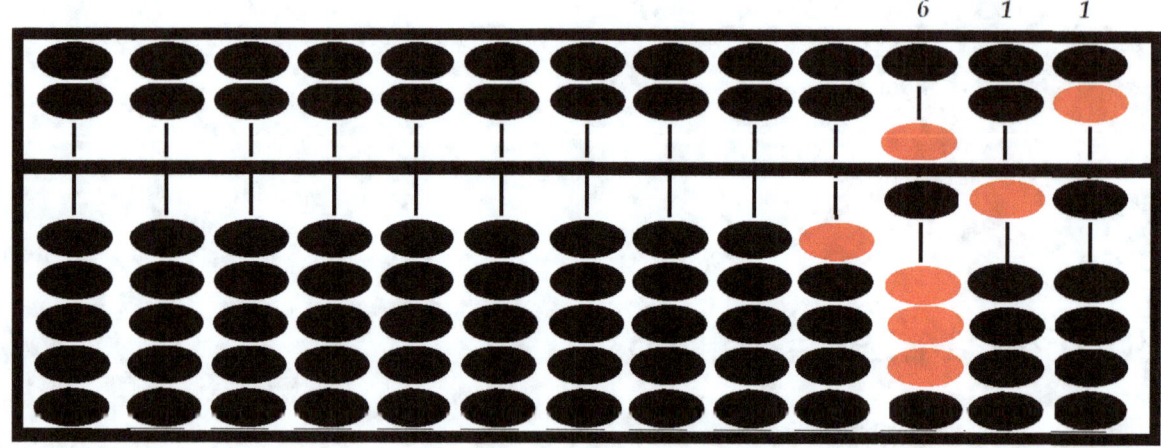

Set 5,070 and then subtract 3,982.

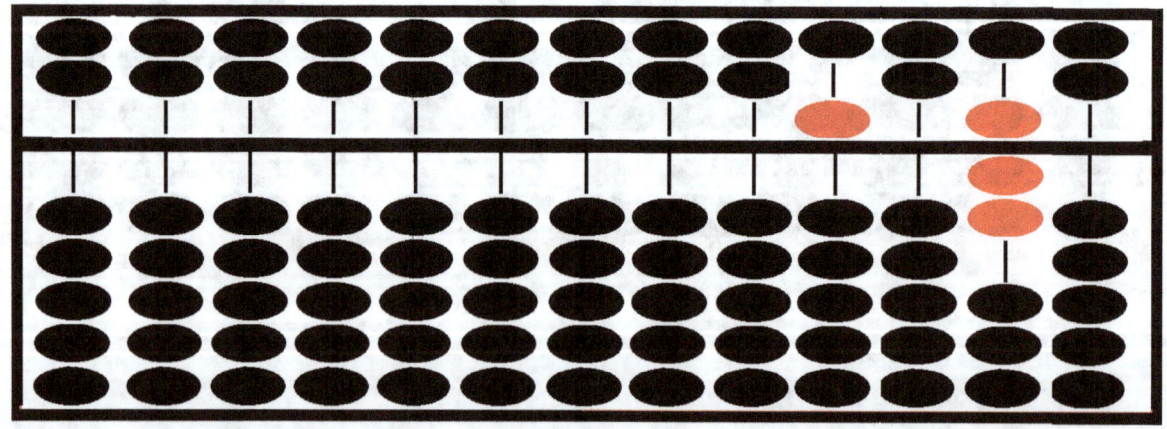

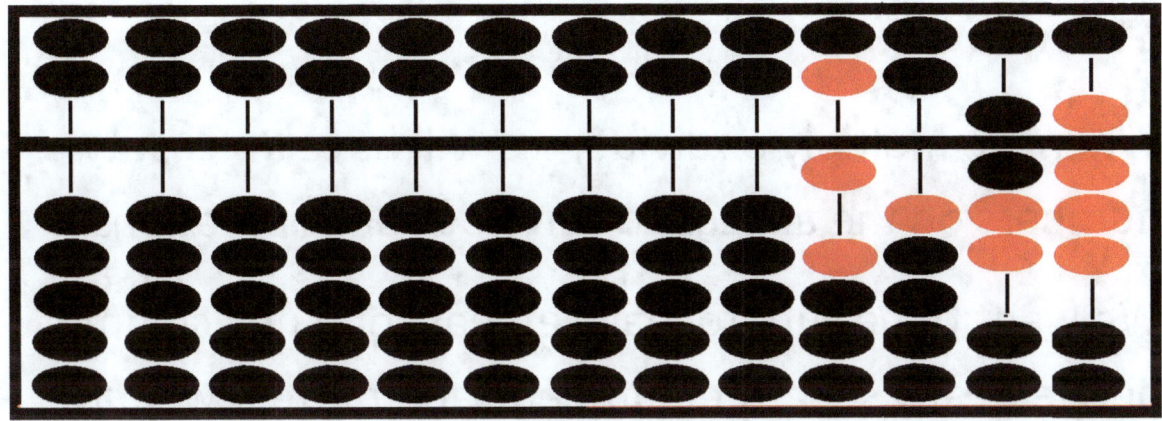

5,070 − 3,982 = 1,088

If you got this one, then you know how to add and subtract on the abacus.

You will be able to represent any number with our 5s and 1s!

Try these exercises:

934 − 29 =   1,845 − 590 =   107 − 99 =   919 − 75 =

31,906 − 2,057 + 1848 +29 − 1005 =

2,340 + 20,340 + 22,044, + 9 − 65 =

Multiplication

When we multiply or divide, we will have at least two numbers on the abacus to carry out our operations.

We want to be sure to leave at least one null column between any separate numbers, and more if we have room, for example, multiplicand, multiplier and product, or divisor, dividend and quotient.

There is always a need for extra caution when working with numbers ending in zero. A column holding a 0 can be mistaken for an empty column. For numbers ending in zero, be sure to leave an extra column.

As we multiply we add numbers to the product from right to left, so start your answer as far right on the abacus as space permits. If you have need for decimal places be sure to leave space on the right. All fraction remainders should be converted to the nearest decimal equivalent.

1 digit multiplier

104 x 4

Set 104 to the left and leave at least one column as a separator between the multiplicand and the multiplier.

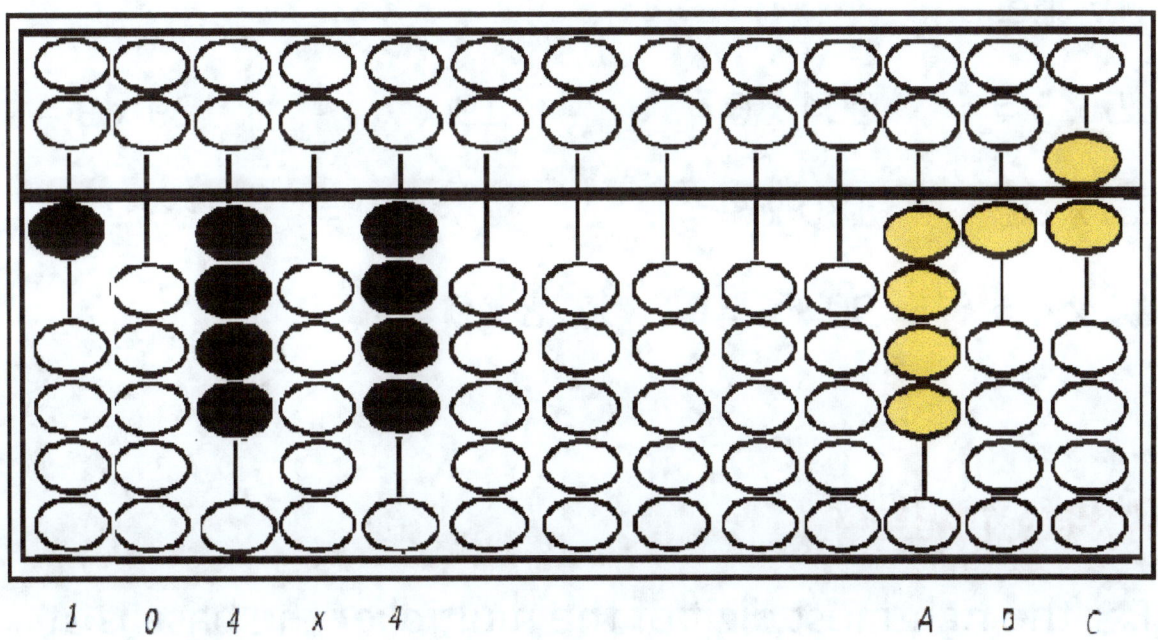

104 x 4 = 416

The answer starts with the base for the 4 at column C and it is 16 (4x4). For the next digit of the multiplicand, the base is at column B, and the product is 0.

(4 x 0), the base then moves to A for the next digit and is 4, (4 x 1).

We started the answer at C because we did not know how many spaces (columns), we would need for the answer.

Try these:

15 x 5 =   204 x 6 =   19 x 7 =   129 x 9 =

12 x 20 =   180 x 3 =   28 x 49 =   16 x 7 =

14 x 4 =   90 x 6 =   228 x 9 =

2 digit multiplier

For the rightmost digit of the multiplier the base is at the rightmost column.

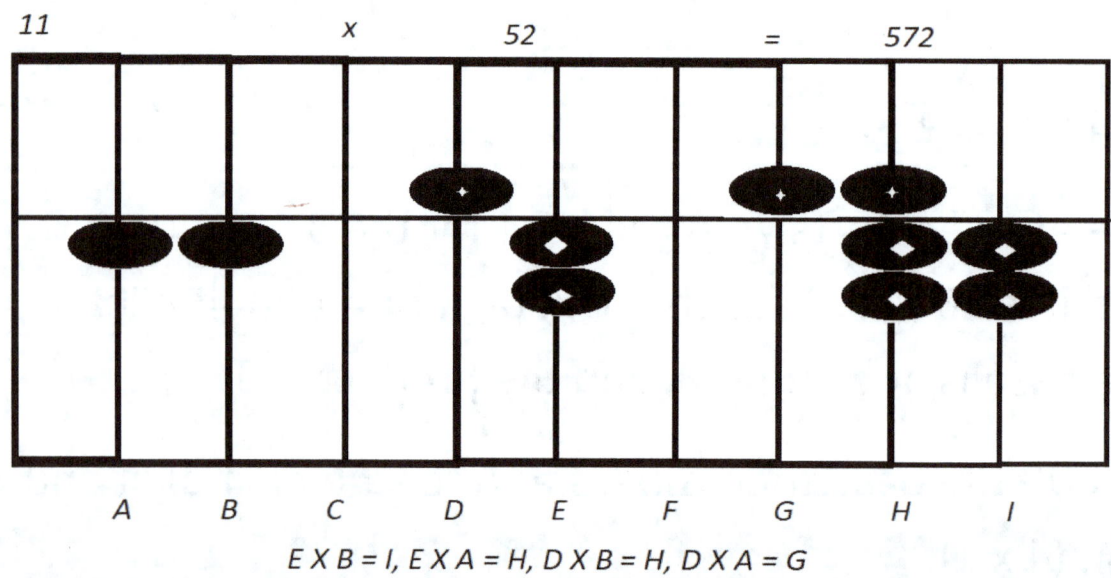

EXB=I, EXA=H, DXB=H, DXA=G

As the digits of the multiplier are applied to the digits of the multiplicand, the starting base is moved one column to the left in the multiplier.

The base is also moved one place to the left when we use the next digit of the multiplicand, going right to left, in both the multiplier and multiplicand.

Set up 213 x 18

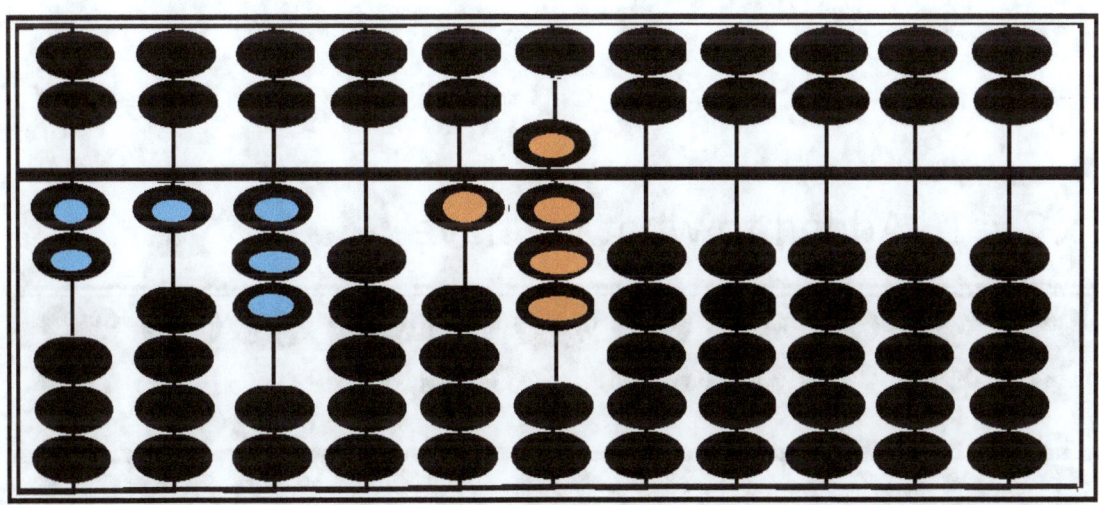

8 x 3 = 24

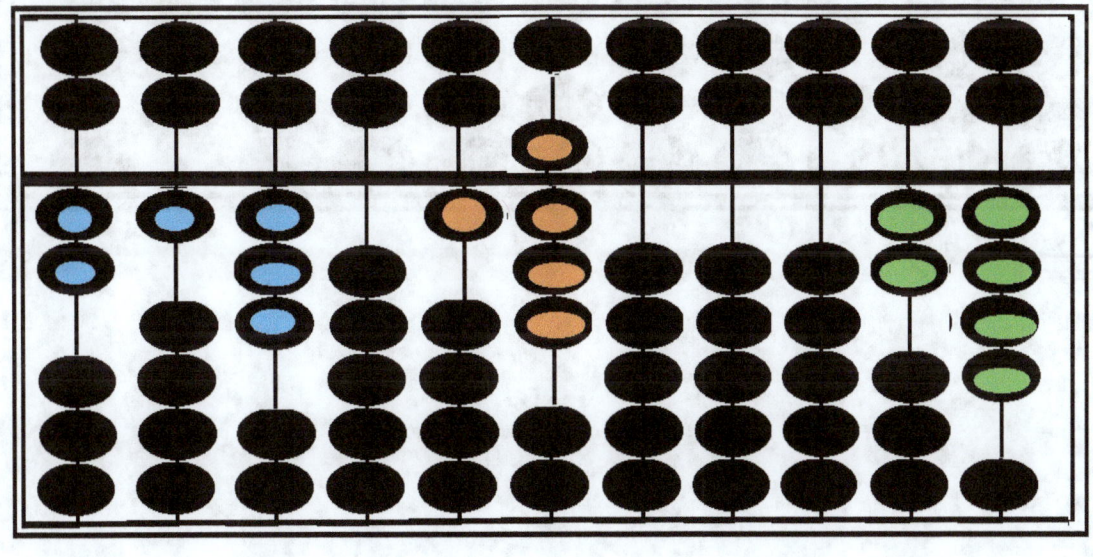

8 x1 = 8. We add 8 to what we have

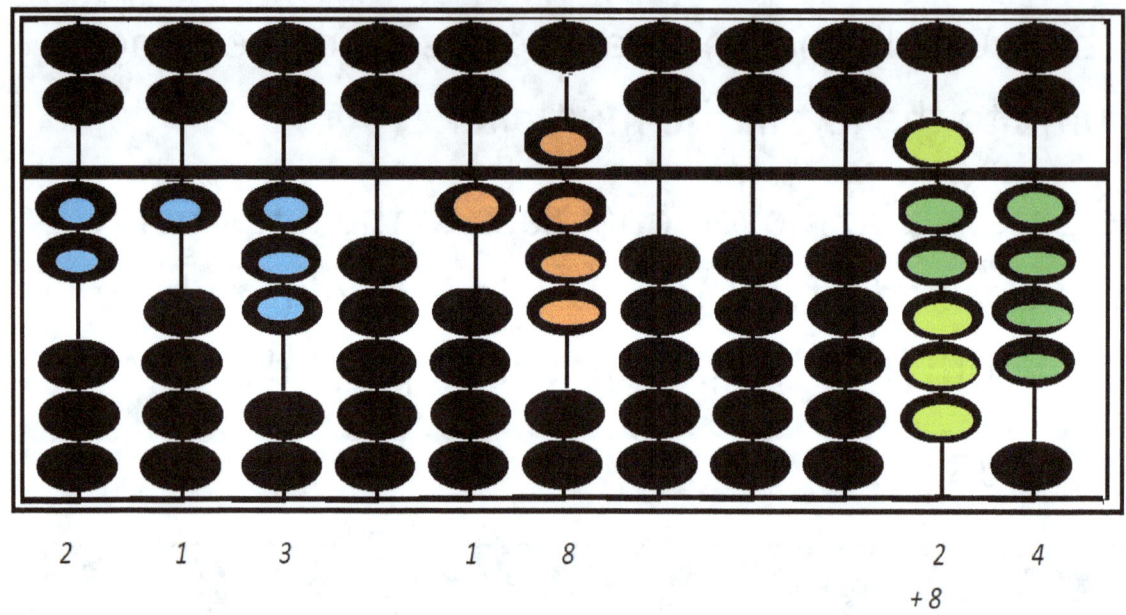

8 x 2 = 16 Added to what we have

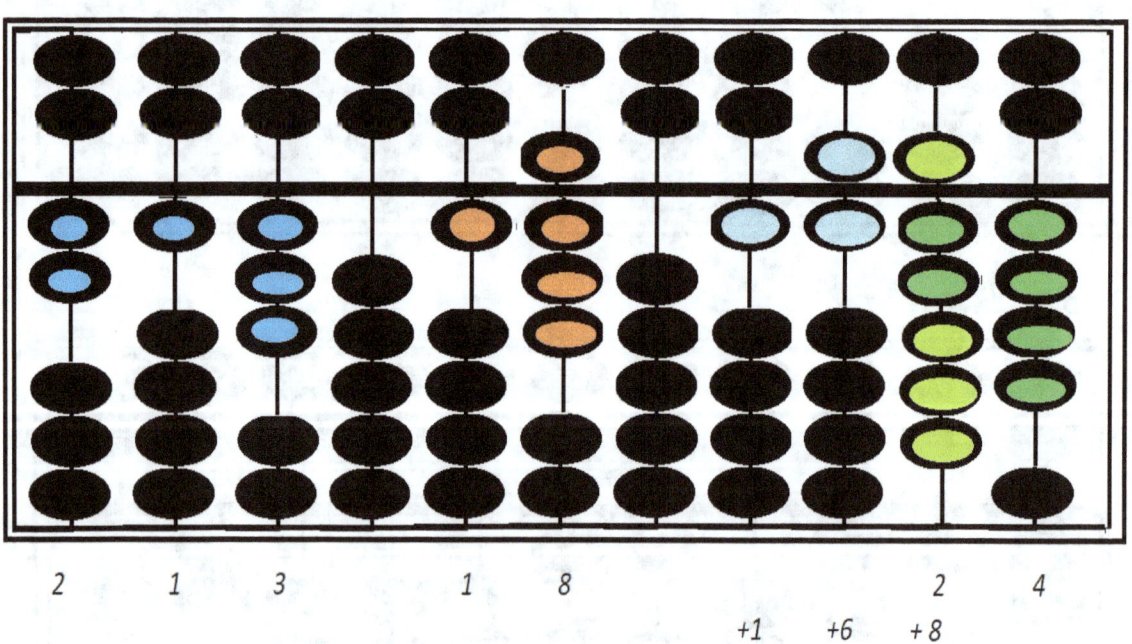

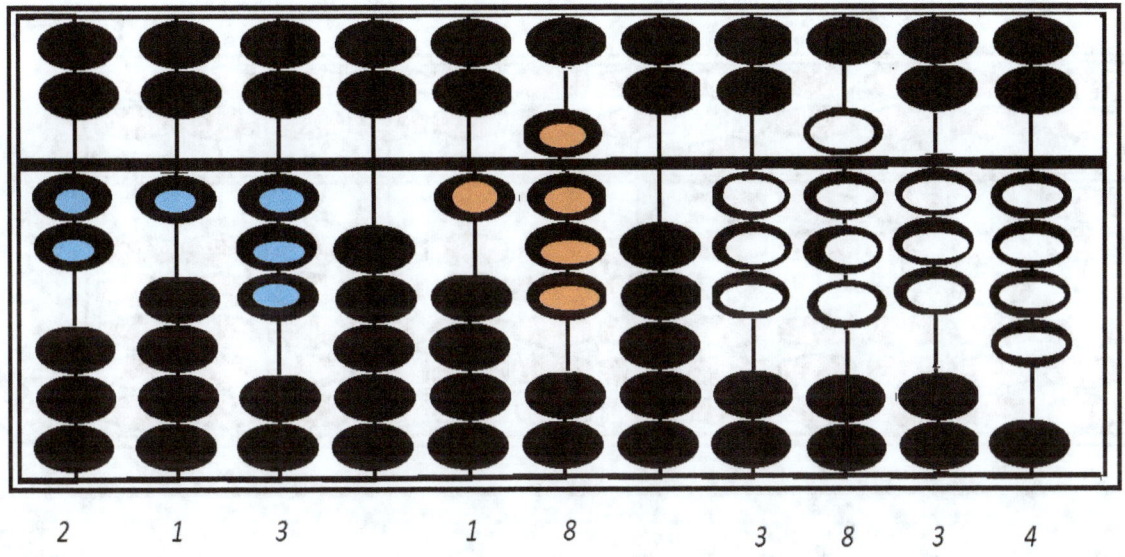

Now the multiplier is the 1 in 18 so the base is at the second column from the right.

1 x 213 = 213, so the solution, when this product is added, is 3,834.

During the calculations we did not forget to adjust when 2 Heaven Beads where at the bar! We did not forget to move an Earth Bead up on the left column when we pushed 2 Heaven Beads on the next right column back to the top! We did not forget to move a Heaven Bead down, on the same column, when we moved 5 Earth Beads down If we did then maybe we have the wrong answer.

3 digit multiplier

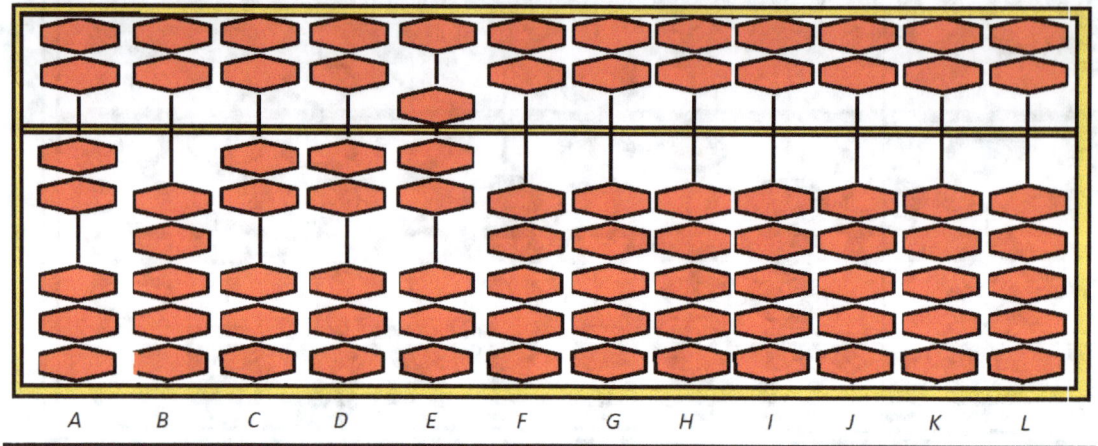

Let us multiply 2 x 227. The base for multiplier 7 on E is at L, for 2 at D the base is at K, and for 2 at C it is J.

7 x 2 = 14 and we set 14 on the right two columns as 1 and 4

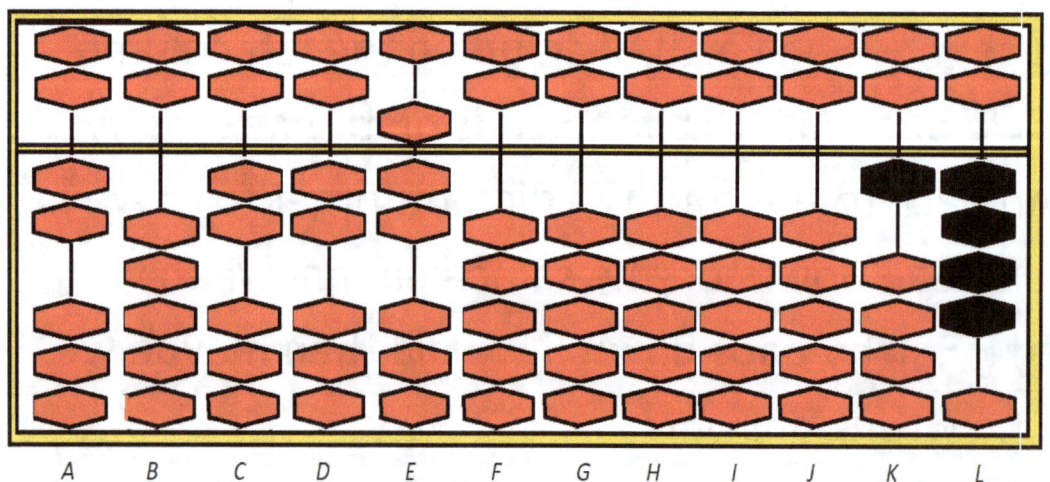

2 x 2 = 4 so we move up 4 Earth Beads on K

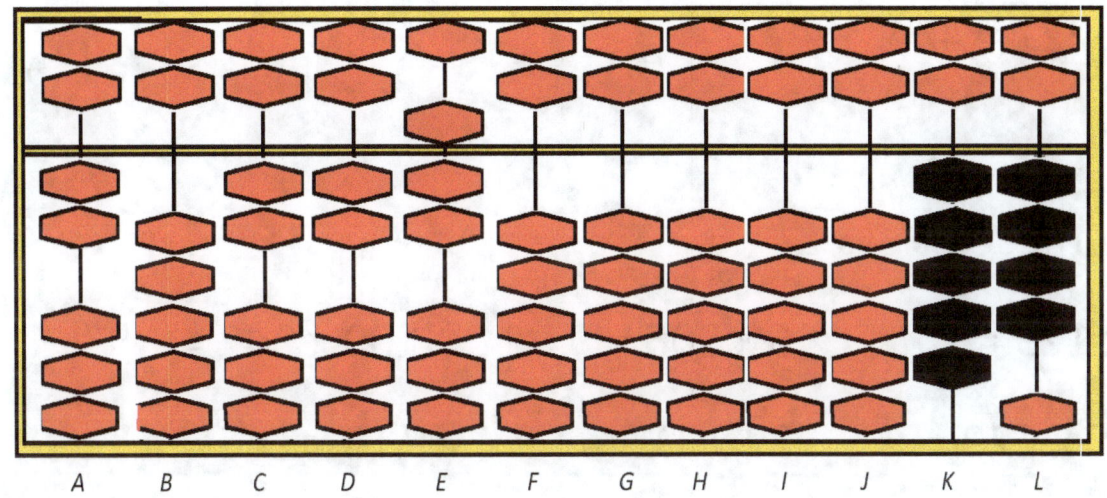

2 x 2 = 4 so now we add 4 Earth Beads on J

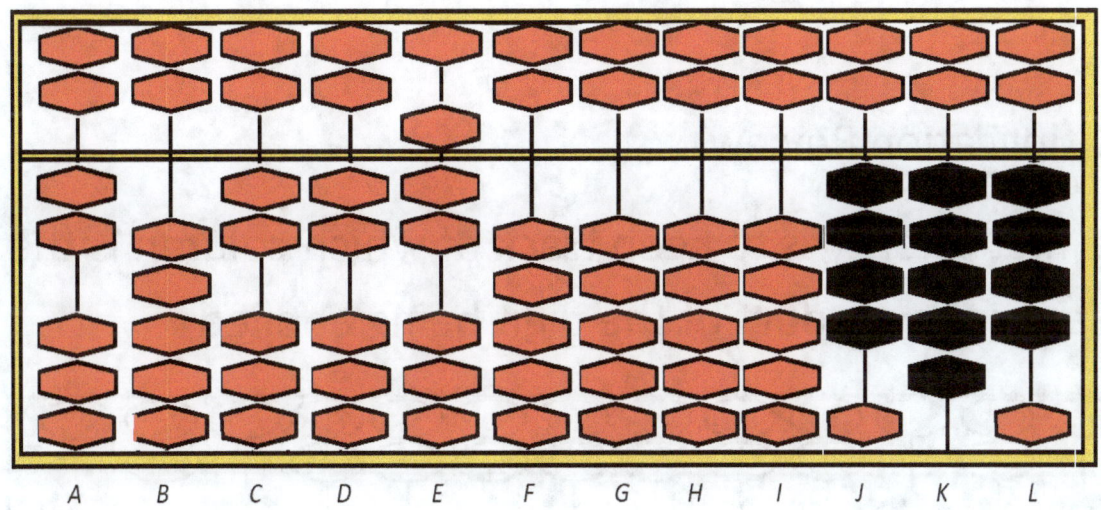

Now we need to adjust for K, all Earth Beads are used

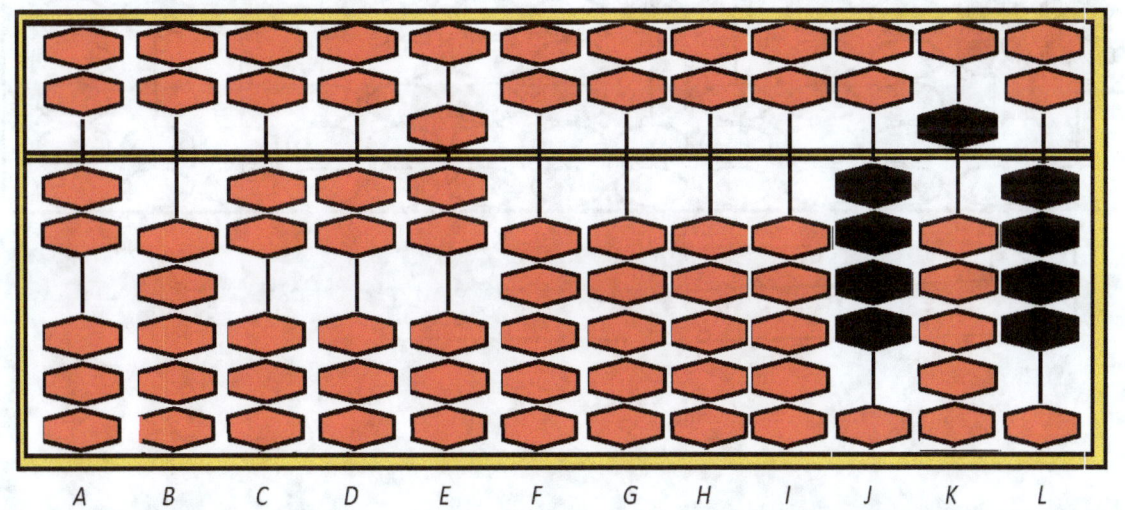

and they are replaced by 1 Heaven Bead, we have

2 x 227 = 454.

Try these..

4 x 129 =   6 x 206 =   109 x 342 =

240 x 120 =   22 x 123 =   118 x 246 =

992 x 183 =

Multiplication Review

The first priority is to keep track of which column is the base for the product of the digit being applied.

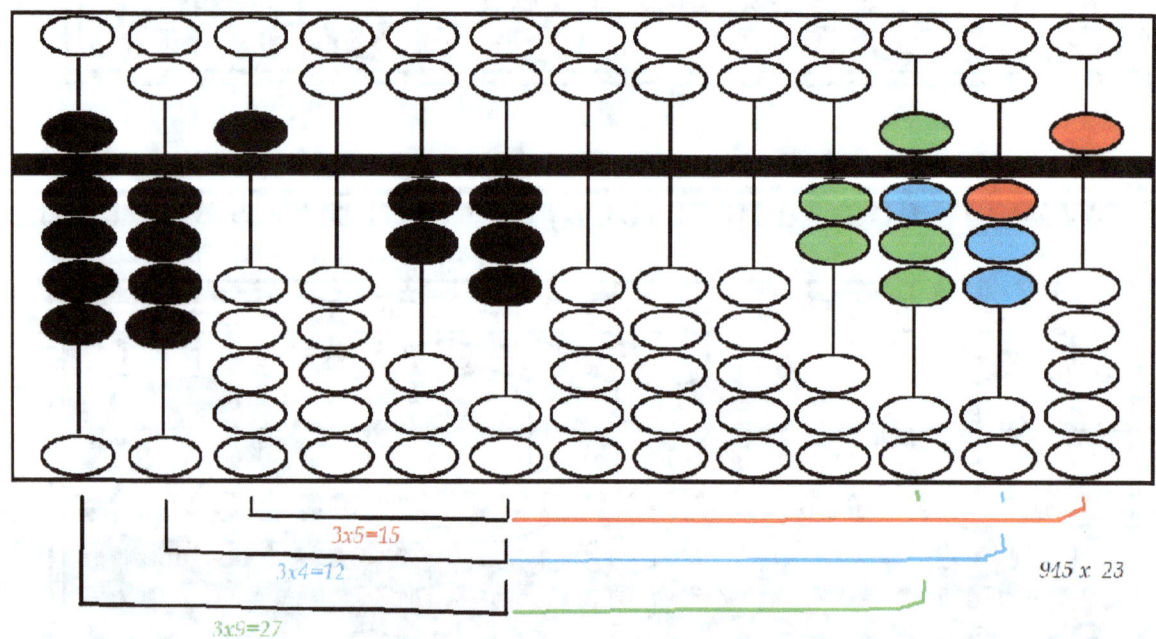

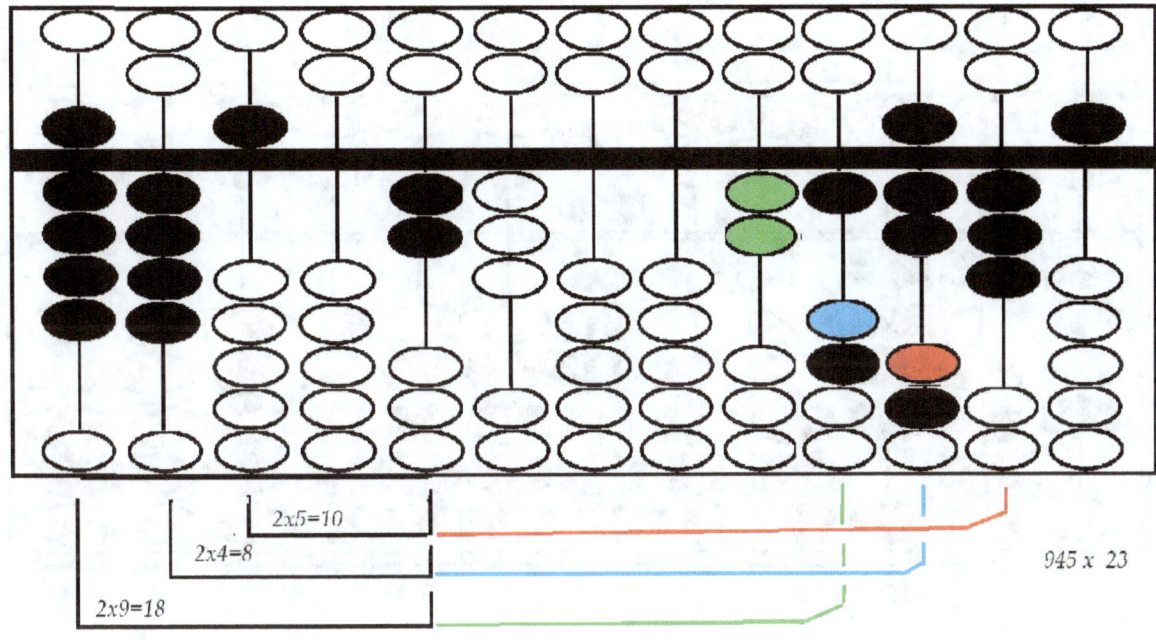

$$945 \times 23 = 21,735$$

Division

When dividing on the abacus we need to allow sufficient spaces for the quotient as we enter it from left to right so we start our quotient at the leftmost column possible.

414 / 18

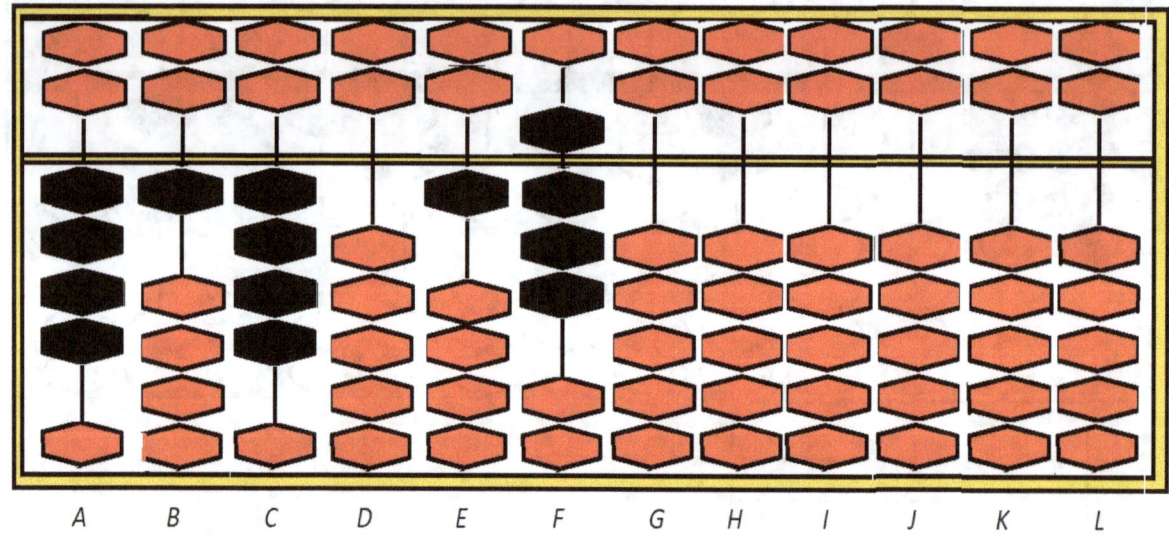

This quotient will start at column I.

18 goes into 41 twice (2 x 18 = 36), 41 − 36 = 5 (at B)

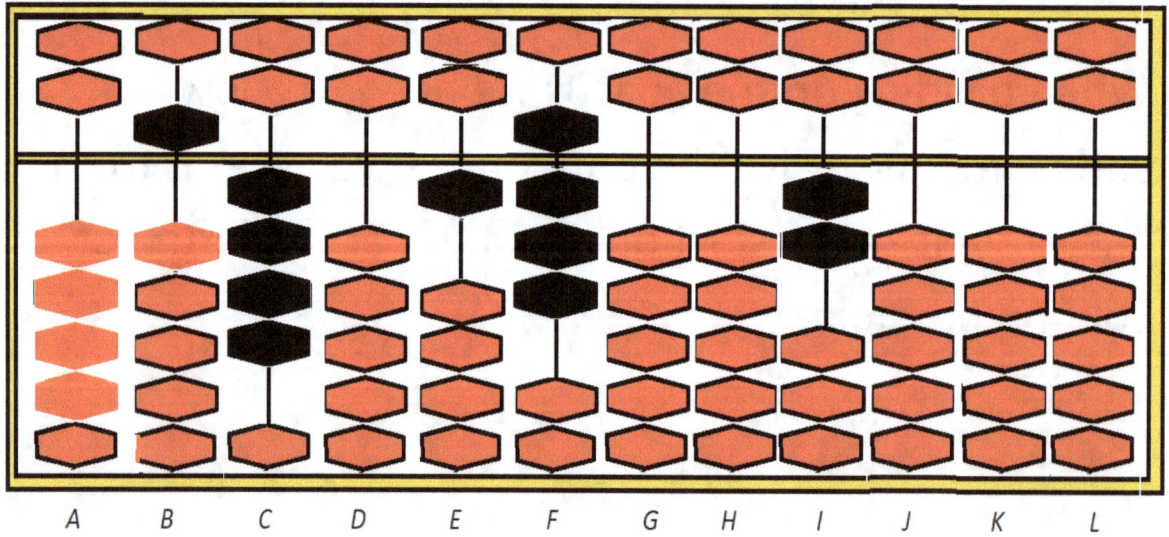

We have reduced the 41 by 36 as you can see on the abacus we now have 54 to divide by 18.

18 goes into 54 three times (3 x 18 = 54).

(54-54 = 0 at C, D)

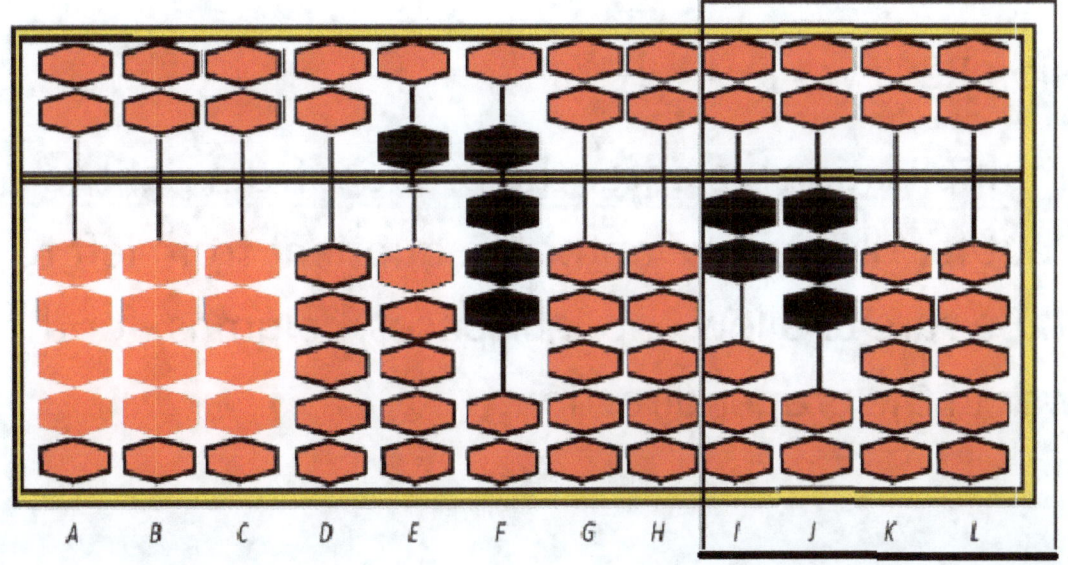

Add the 3 Earth Beads to J. Subtracting 54 leaves us no remainder with 23 as the quotient. We started our answer at column I, leaving 2 decimal places for a possible remainder.

We can also find the quotient by repeatedly subtracting 18 from 414; we would do it 23 times.

Can you complete these?

225 / 25 =   216 / 3 =   608 / 19 =

135 / 15 =   658 / 72 =   420 / 35 =   57,015 / 105=

Division Review

For multiplication we enter the product from right to left, but for division we enter the quotient from left to right. Be sure to allow enough spaces, columns, for the answer in the proper direction.

(140 x 8) / 5   Remember to allow for the 0.

140 x 8

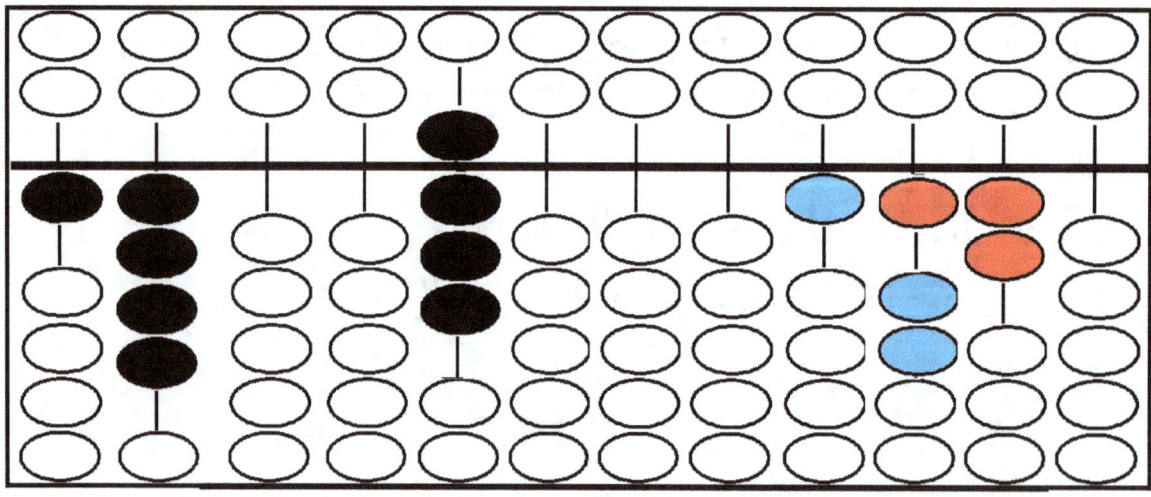

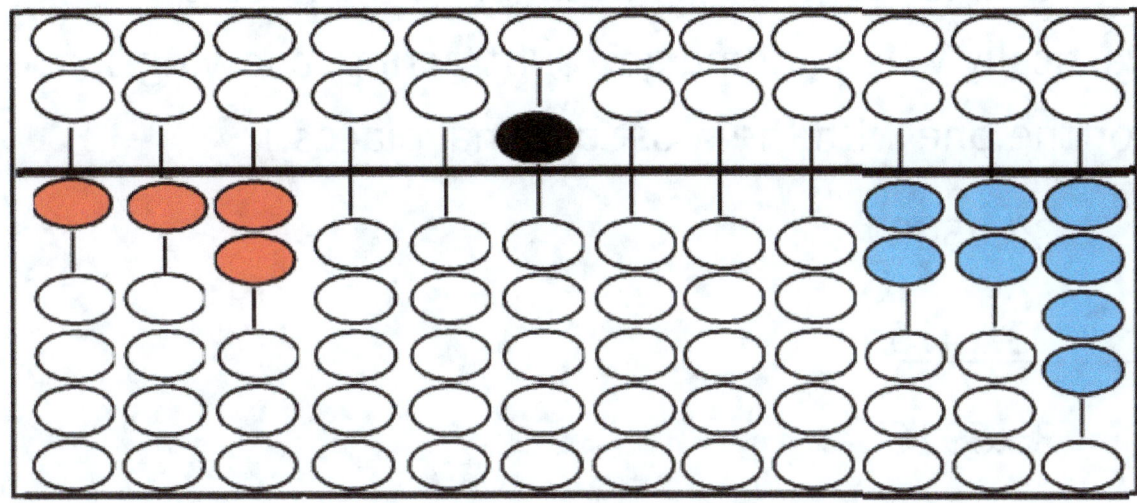

(1,120)/5

224

Decimals

On the abacus the decimal point is wherever you designate. In the previous calculation we might have left some columns on the right for any remainder we might have. On the abacus we do not record fractions in any form but decimals. When there is a fraction in a calculation or result, we convert it to the decimal form. The number of decimal places you will need is determined by how accurate your number must be. Most domestic calculations can do with two decimal places.

You may have more than one number with decimal points on the abacus at the same time.

32.45 + 3.1416 (keep all decimal points lined up vertically when adding and subtracting, allowing space for the one with the most decimal places.)

32.4500

+   03.1416

35.5916

We need to plan for at least 4 decimal places here!

Set 32.4500

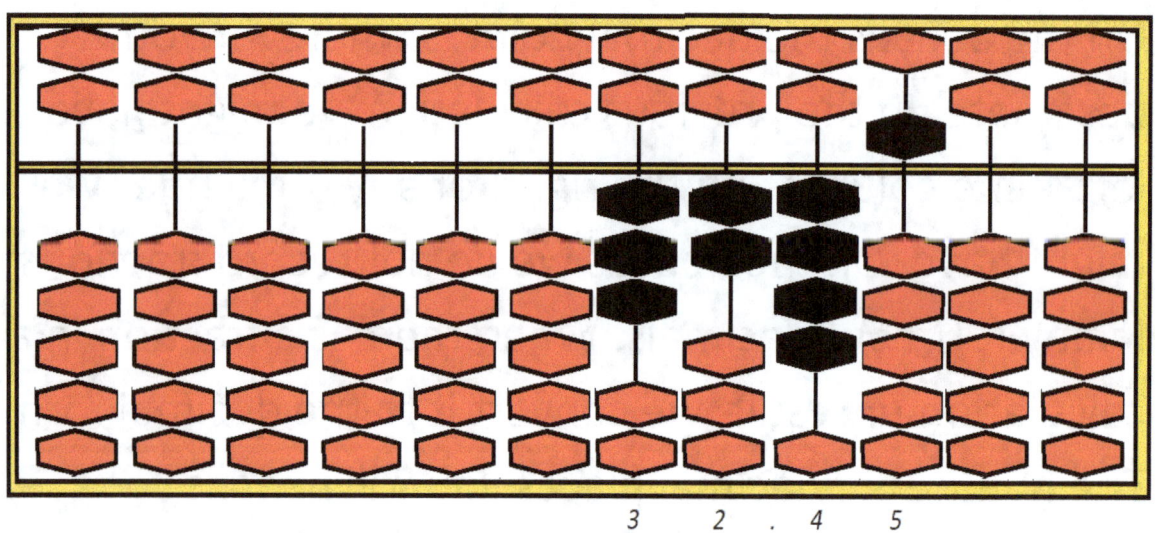

3   2 . 4   5

+ 3.1416

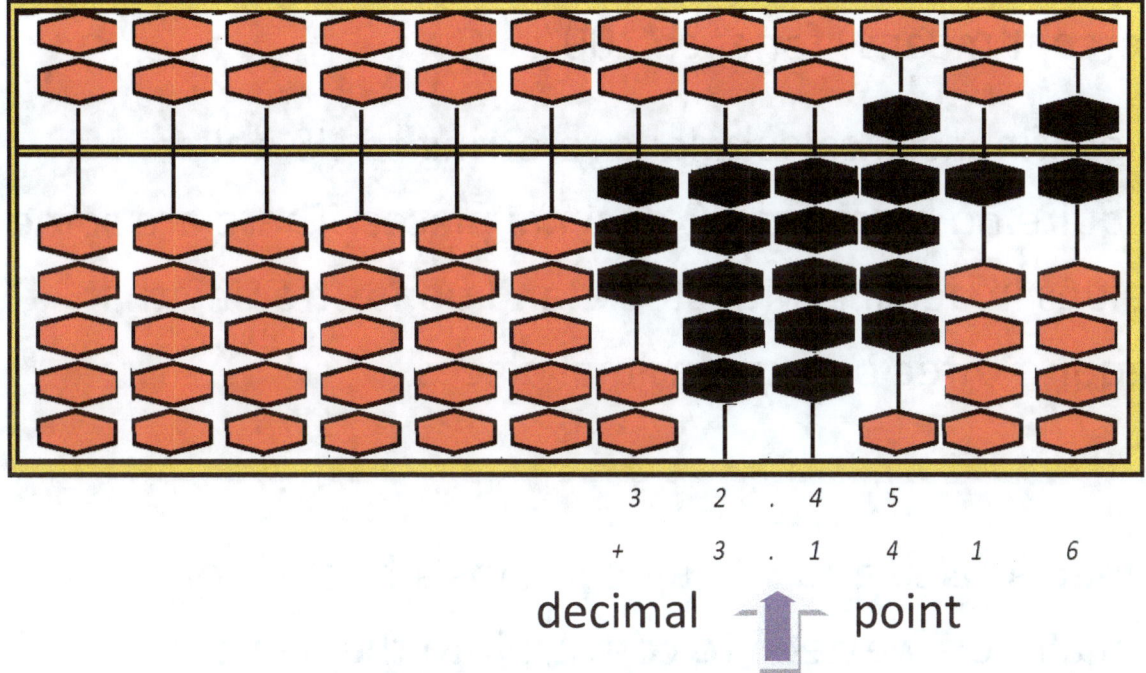

3 2 . 4 5
+ 3 . 1 4 1 6

decimal ⬆ point

Replacing 5 Earth Beads with 1 Heaven Bead:

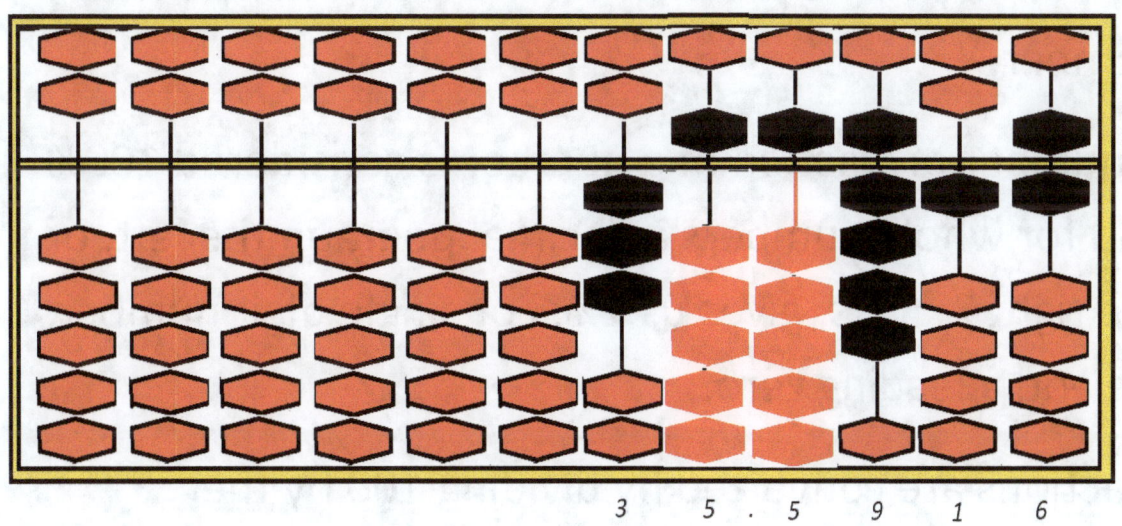

3 5 . 5 9 1 6

You can read 35 point 5916.

Percents

Percent means "for each 100".

Most financial calculations referring to US dollars require no more than 2 decimal places. There are some values in dealing with stocks and bonds trading that require three or four decimal places.

When a calculation result contains a fraction or a remainder, we need to convert it to the nearest decimal equivalent in order to work with it on the abacus.

A decimal point is assumed at the right of any whole number.

A number or fraction of a number is considered 100% and for whole numbers a decimal point on the left, of the right 2 digits, gives us 1%. For single digit numbers we add a leading zero.

Fractions are converted by dividing 100 by the denominator and multiplying the numerator by that quotient.

2/25, 100/25 = 4, 4 x 2 = 8: 2/25 = 8% or .08

What is 5/20 x 6/24? .25 x .25 = .0625 (total decimal places in factors is 4 so we allow for 4 places in the product.)

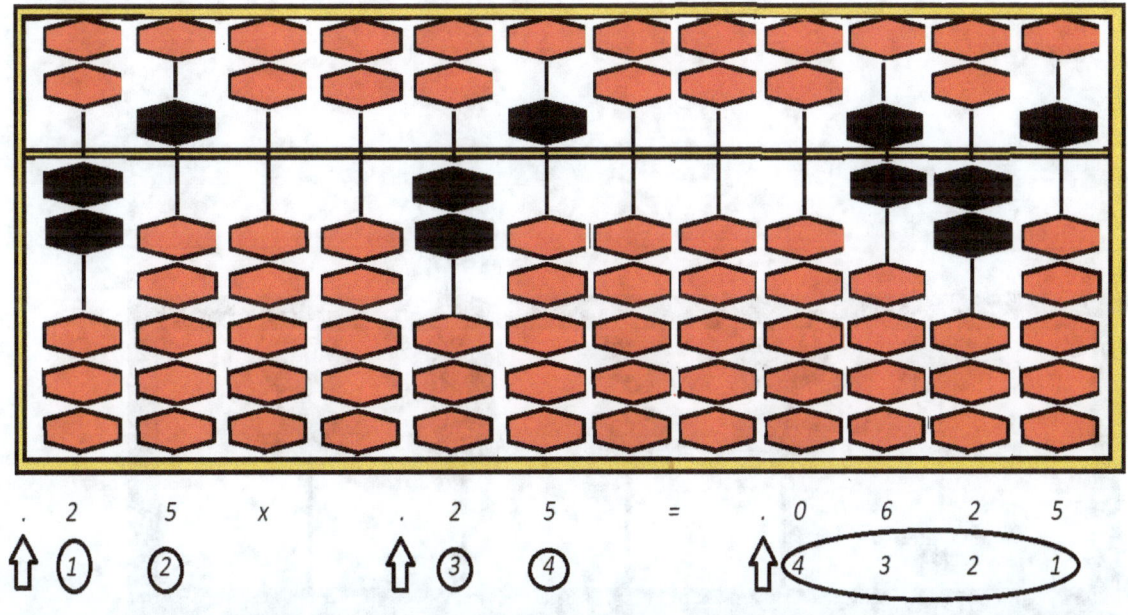

Please total these:

100-23 + 840- 19 -8 + 223- 15 +230 -102 =

(12/3 + 22) x ((6 + 14) x 15))-195 + (234/ 468) =

There are at this time several Abacus tests and certifications available online, in classes or by mail. There is no ultimate authority for certifying Abacus

Expertise in the United States. A certificate from one school or instructor is as good as any other.

You will have earned your certificate for this course by successful completion.

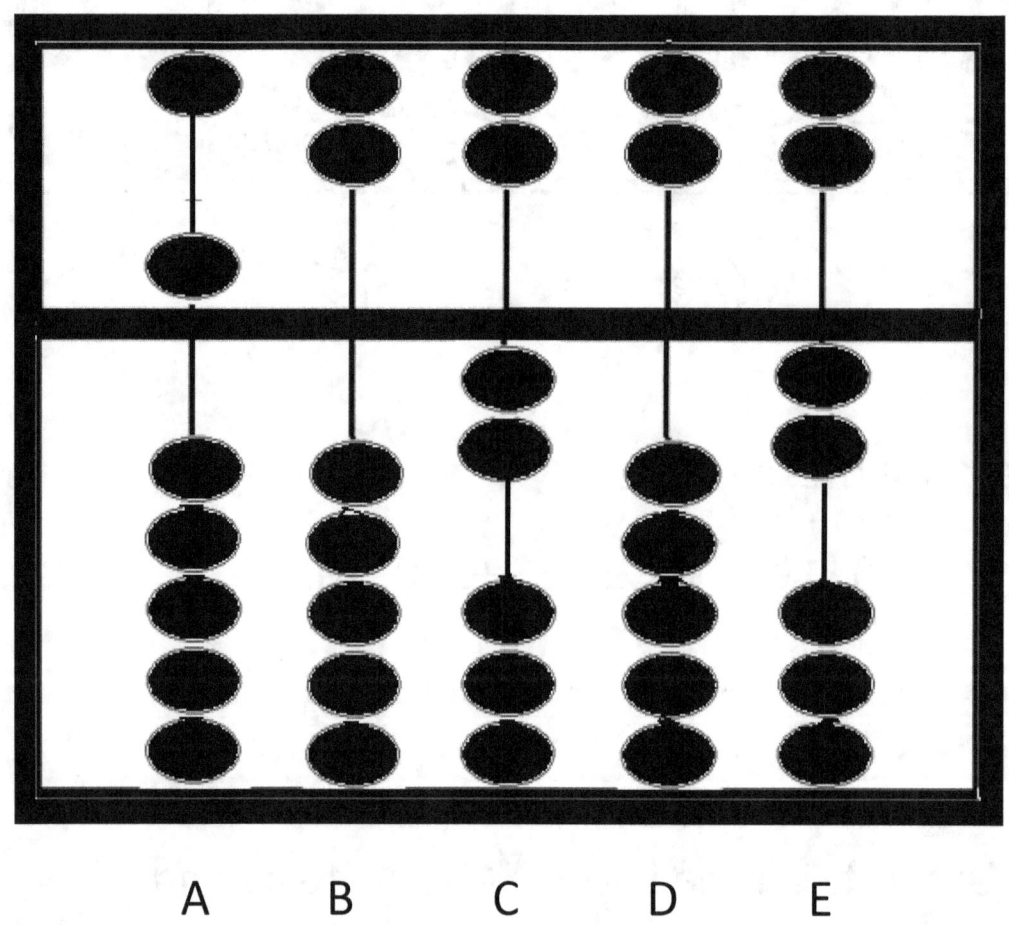

A   B   C   D   E

The Decimal point is between C and D

What do you read the above number to be?

10202_____   502.02_____   5.202_____   .5202_____

a   b   What do we do next?

Move all beads on a up_____

Move a Heaven Bead down and
all Earth Beads up on a _____

Move down all beads on b____

A   B   Compare A and B

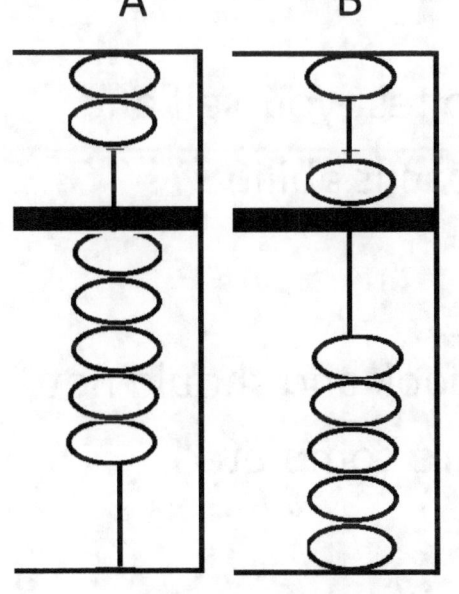

A = B_____

A < B _____

A > B _____

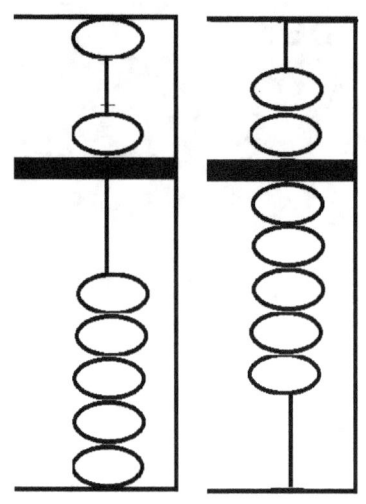

Which comes first and gets adjusted to the other?

C, on the left_____

D, on the right_____

C then D   _____

Neither   ____

D then C   _____

C then D then D_____

The following is a chance for you to test yourself and determine your level of Chinese Abacus ability.

Answers will be at the end of this book and should not be read until all parts of the tests are completed.

Test A

Draw the answer beads positions from your abacus.

a. 22 + 34

b. 16+26

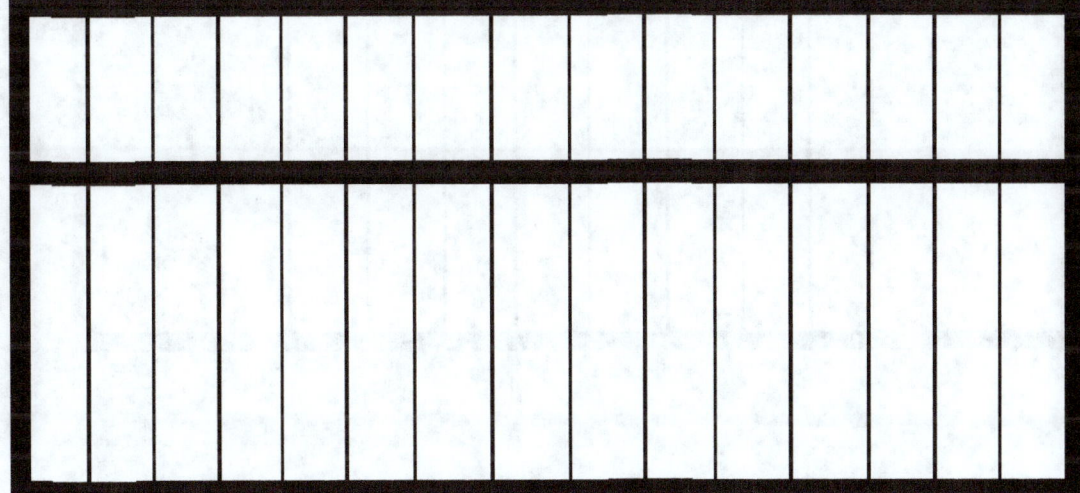

c. 104+220

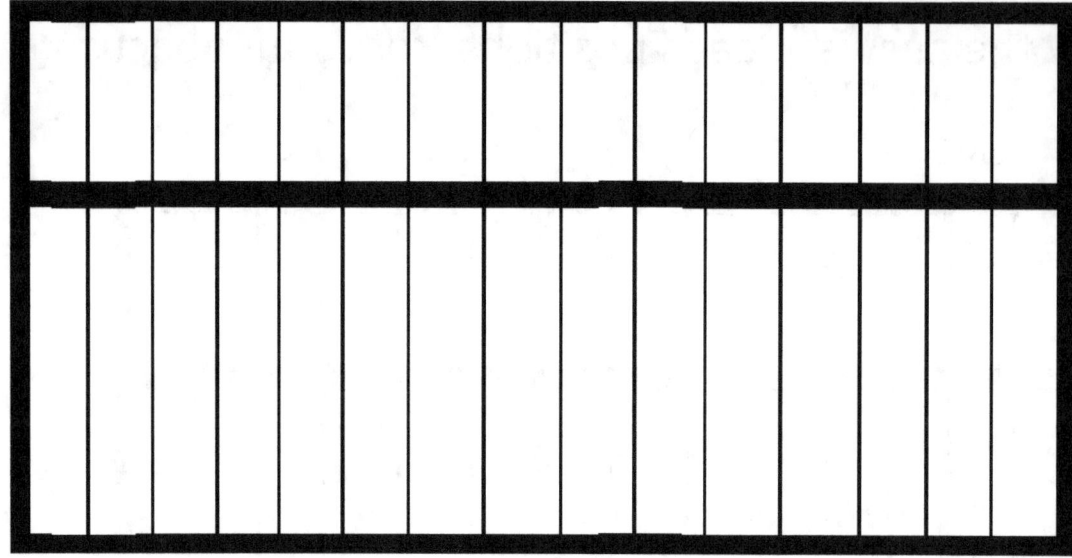

d. 30+70

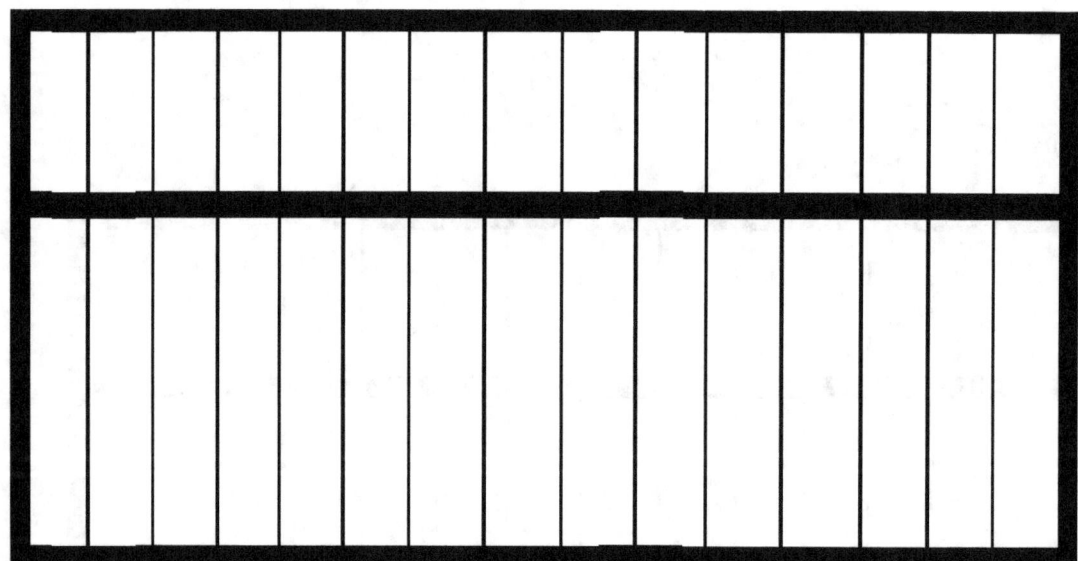

e. 4004+2030

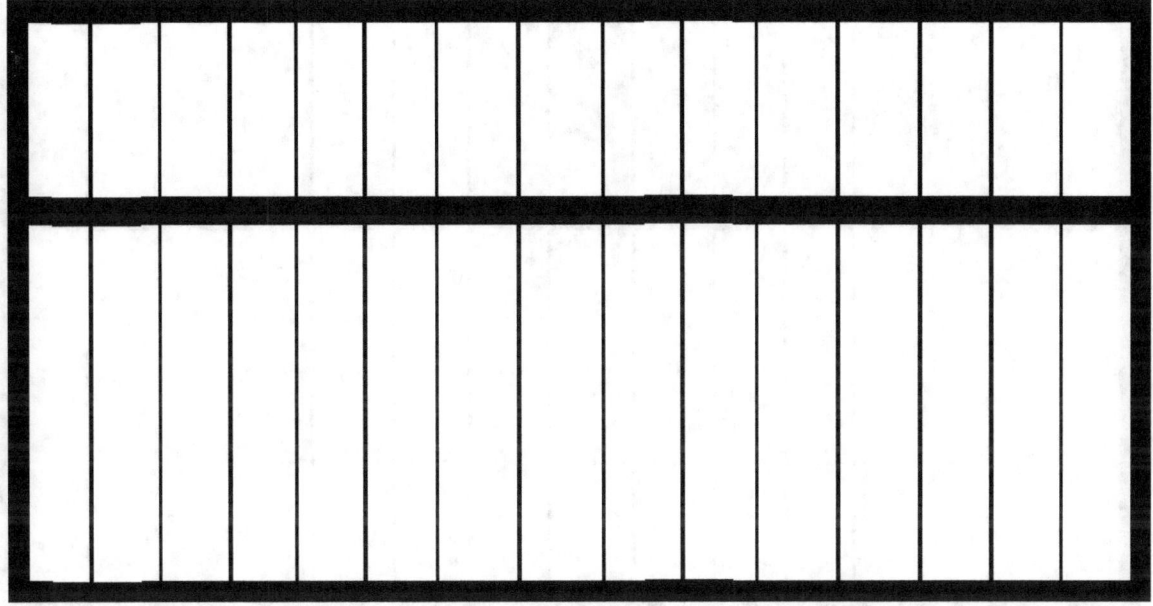

f. 180+810

g. 999+499

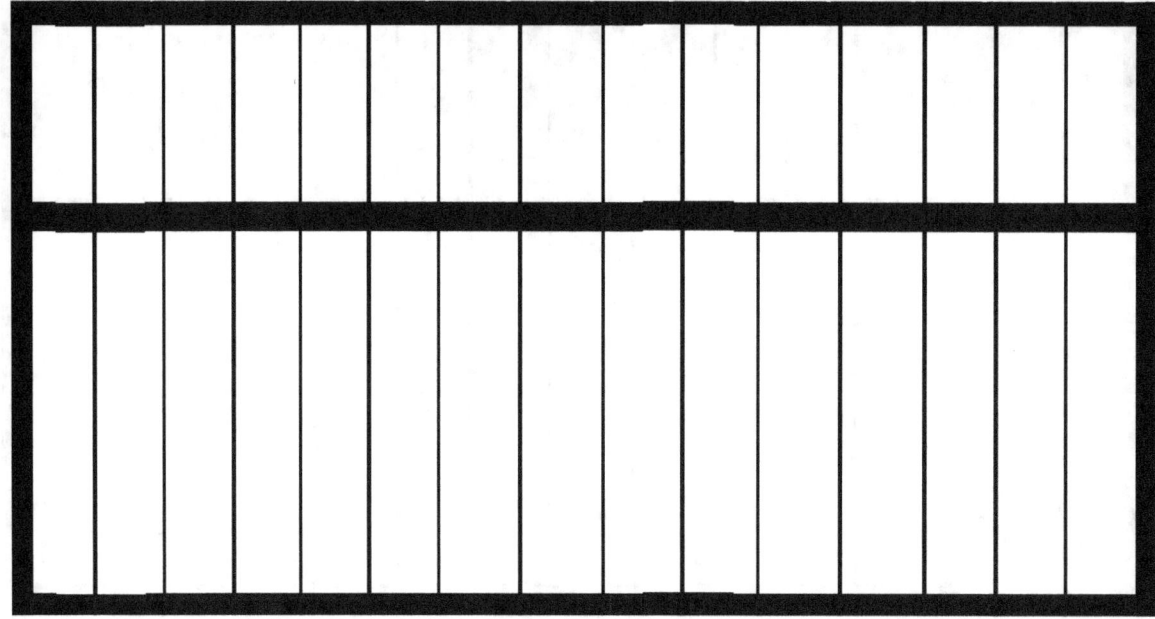

h. 115+15

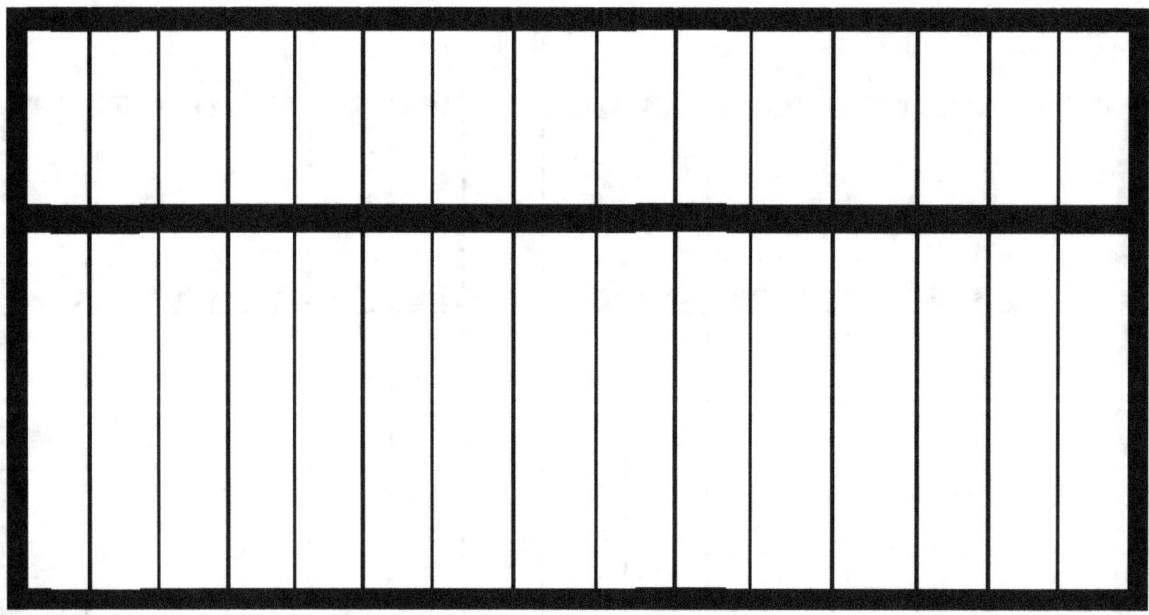

i. 203+219

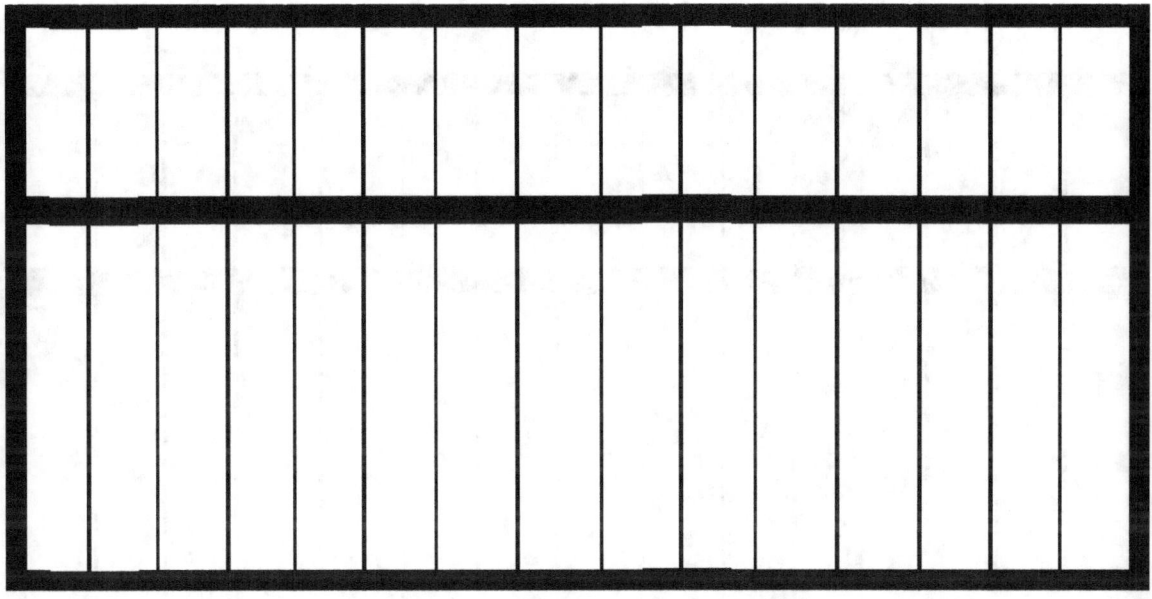

j. 4009+9090

Test B, Diagram beads of result from your abacus

a. 35-18

b. 355-28

c. 298-120

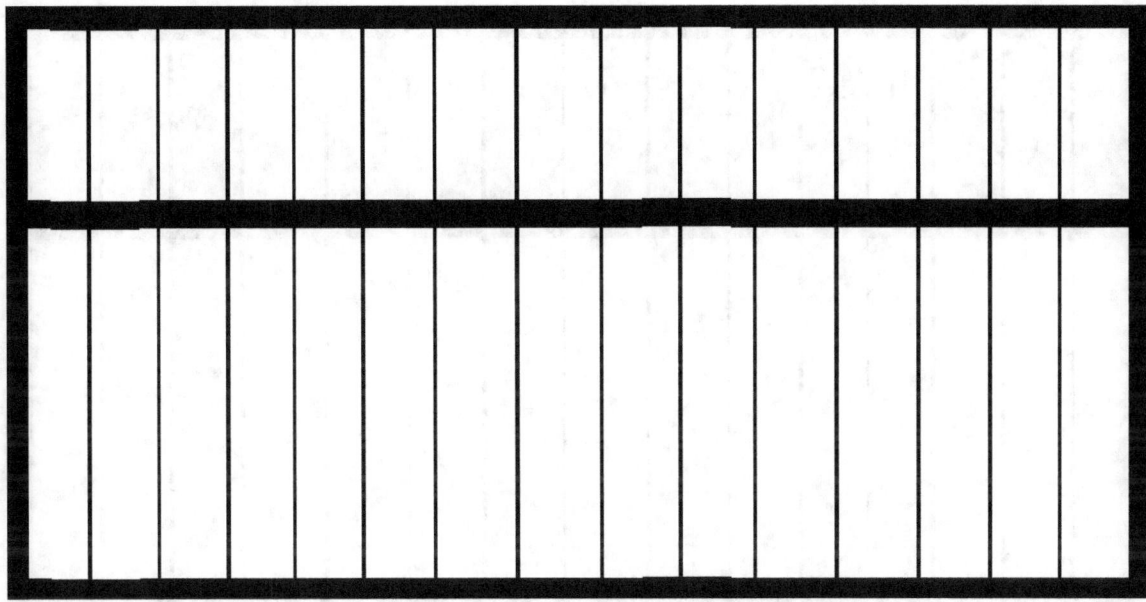

d. 507-203

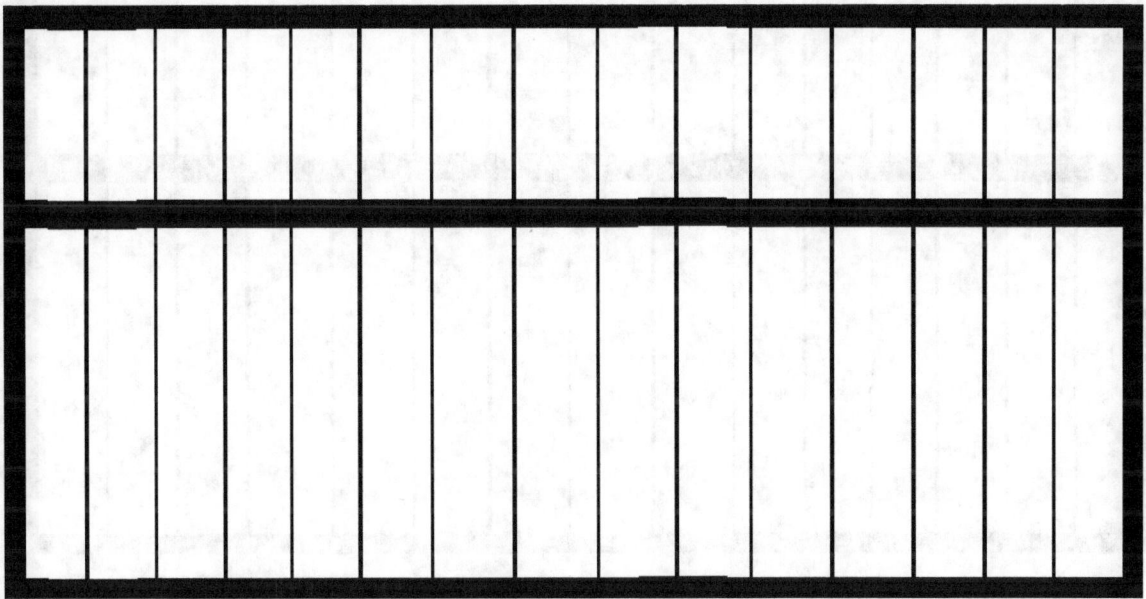

e. 3001-2003

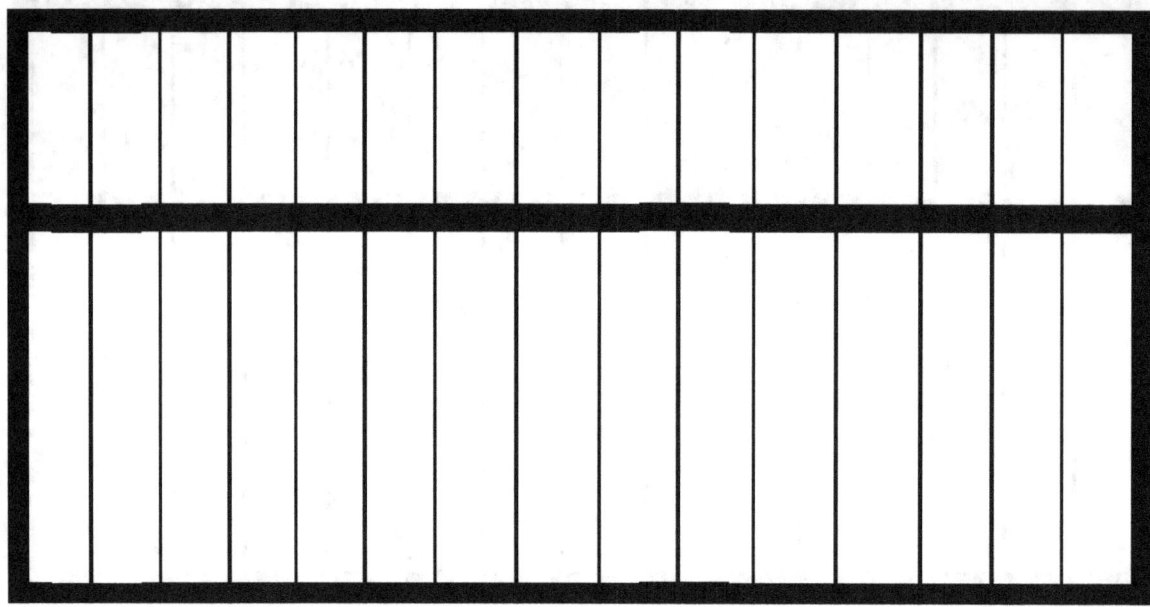

f. 100-97

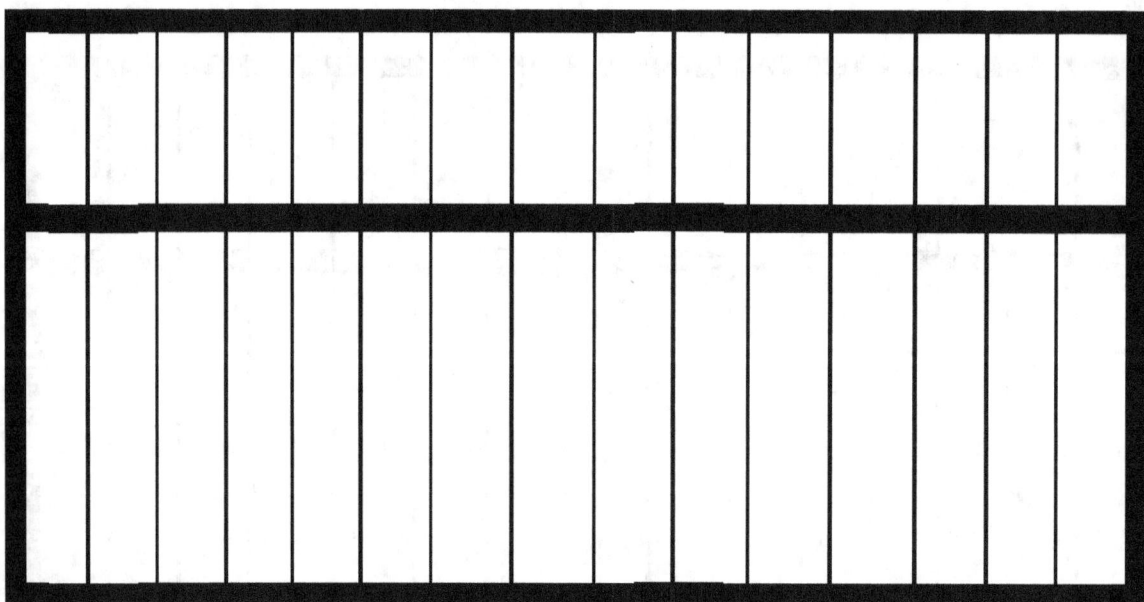

g. 222-199

h. 606-284

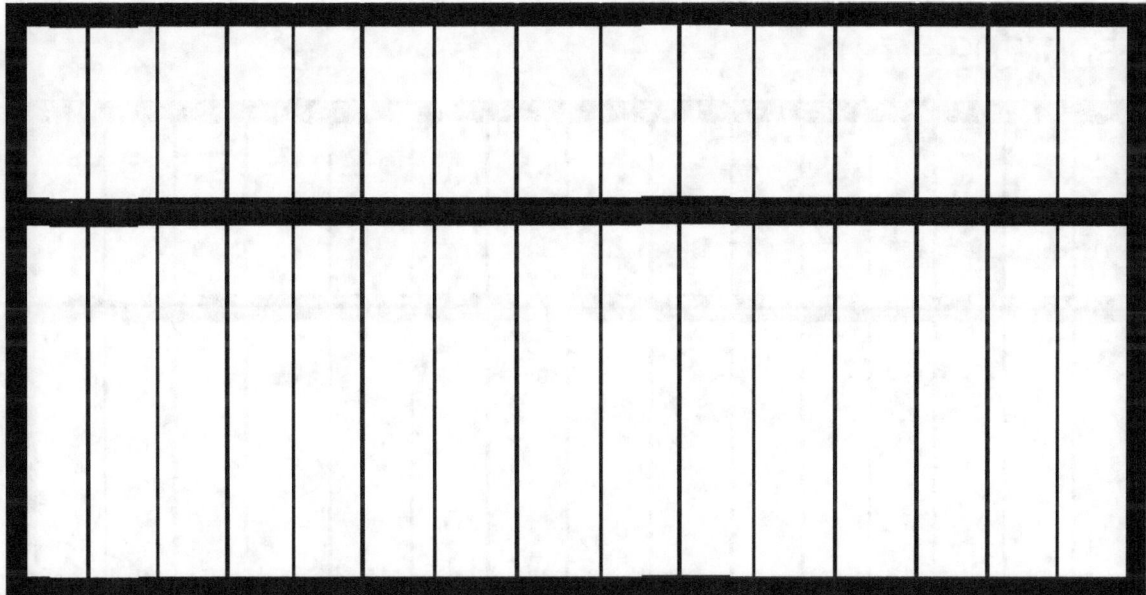

i. 903-605

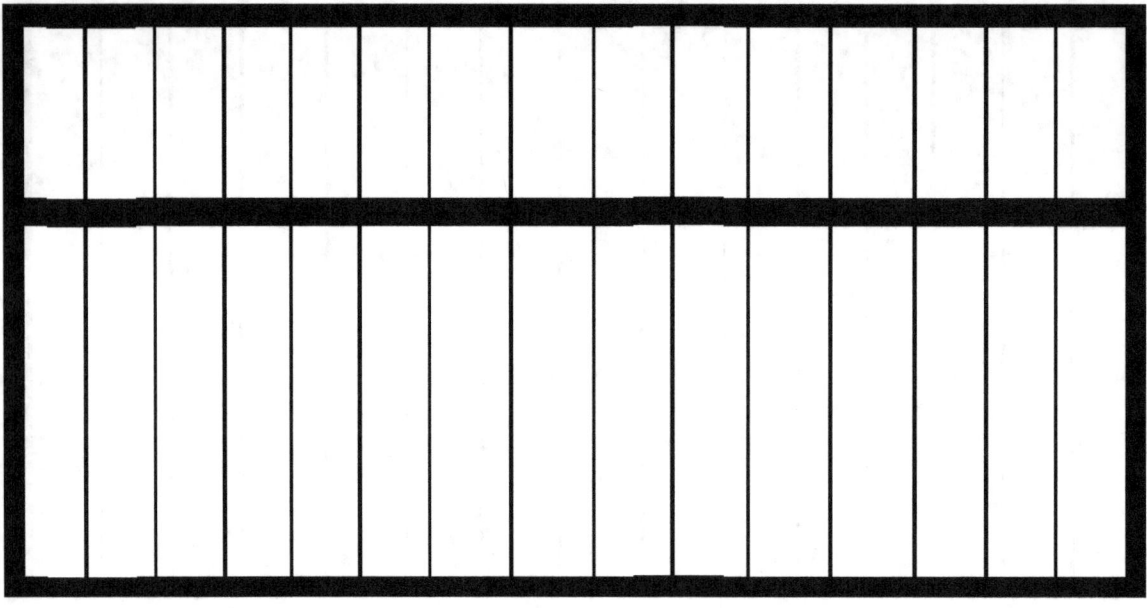

j. 505-209

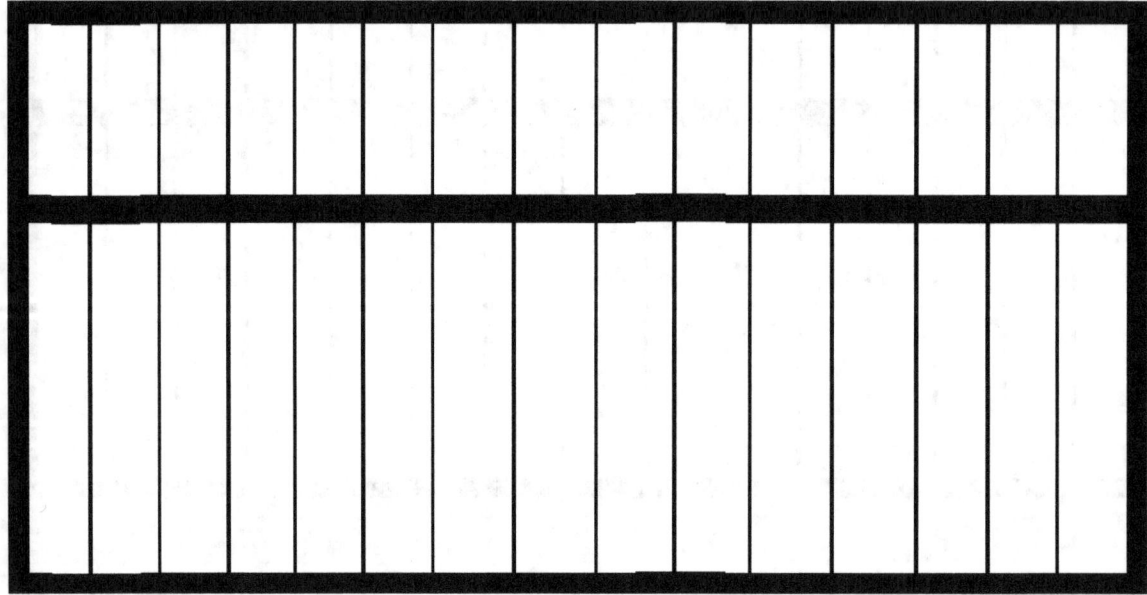

Test C, Diagram solution calculated on your abacus

a. 23+18-9+17

b. 304-210+190-12

c. 213-31+180

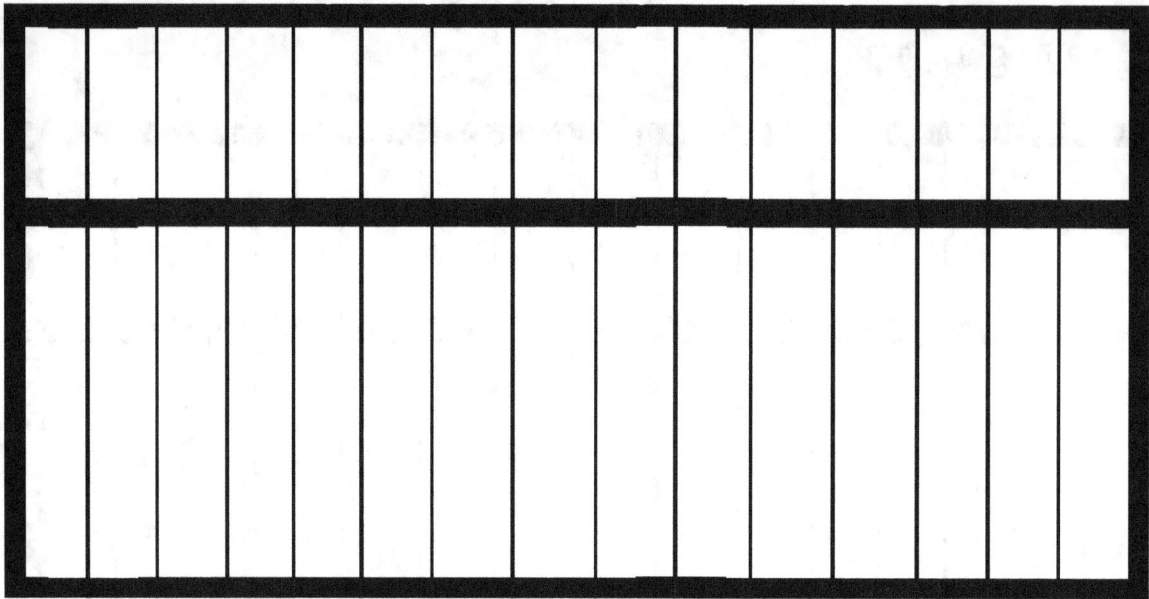

d. 108-9+19-26

e. 999-18+54-22+75

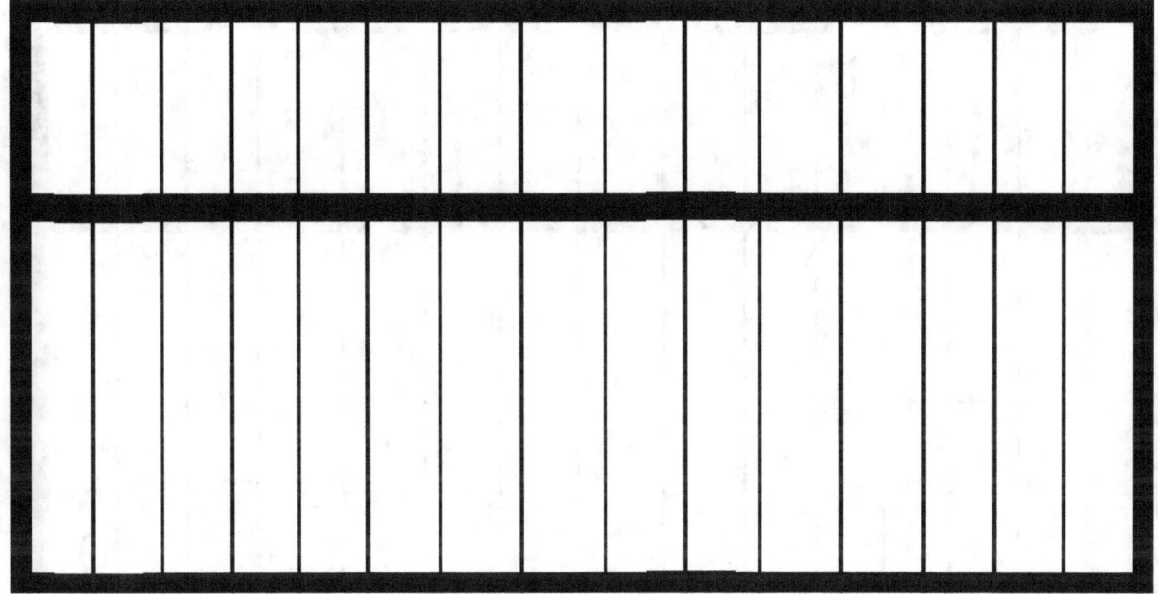

f. 211+112-121+212-2

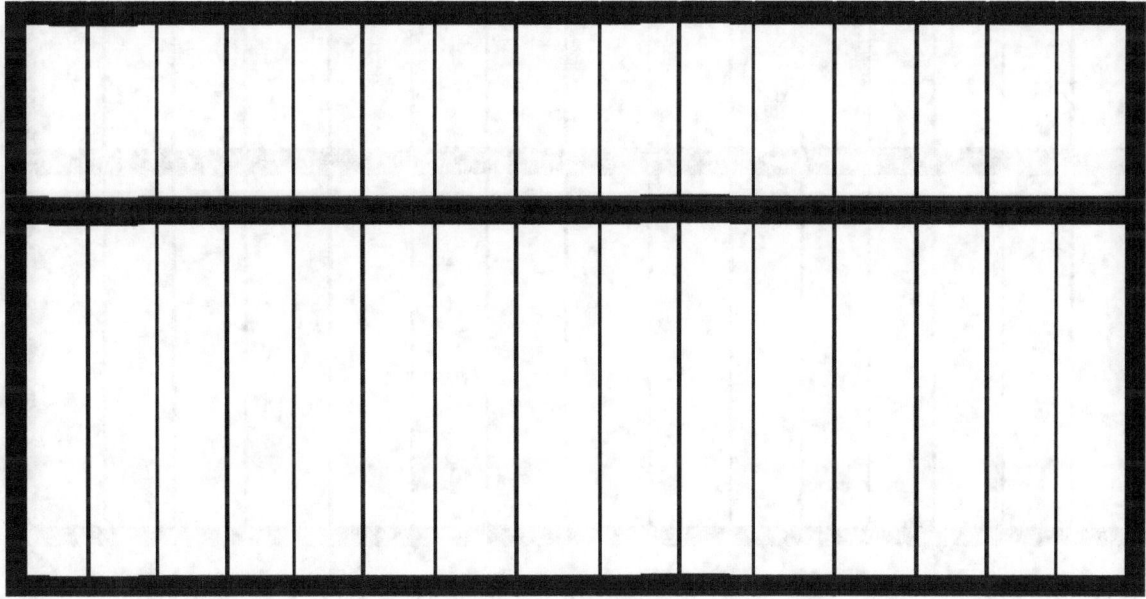

g. 309-27+91-100

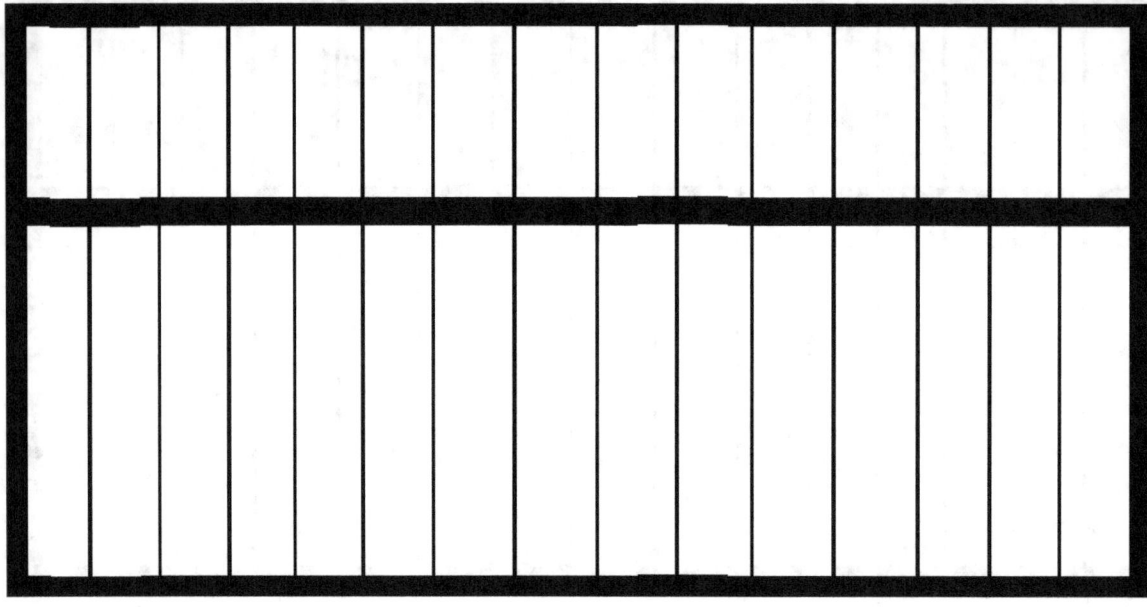

h. 209-87+91-17+20

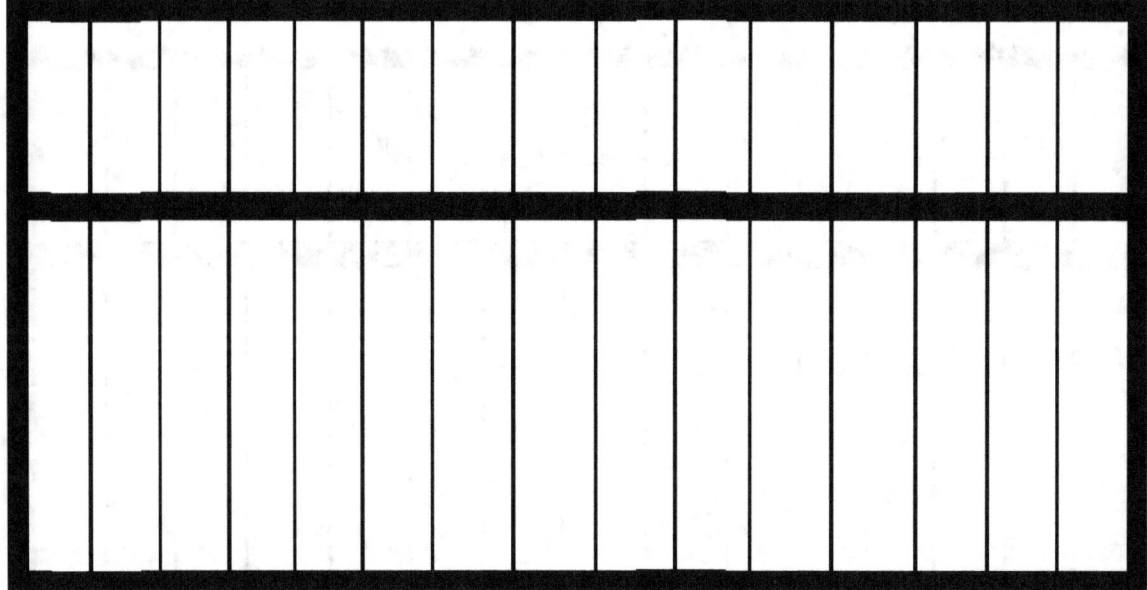

i. 120-19+109-15-22+390

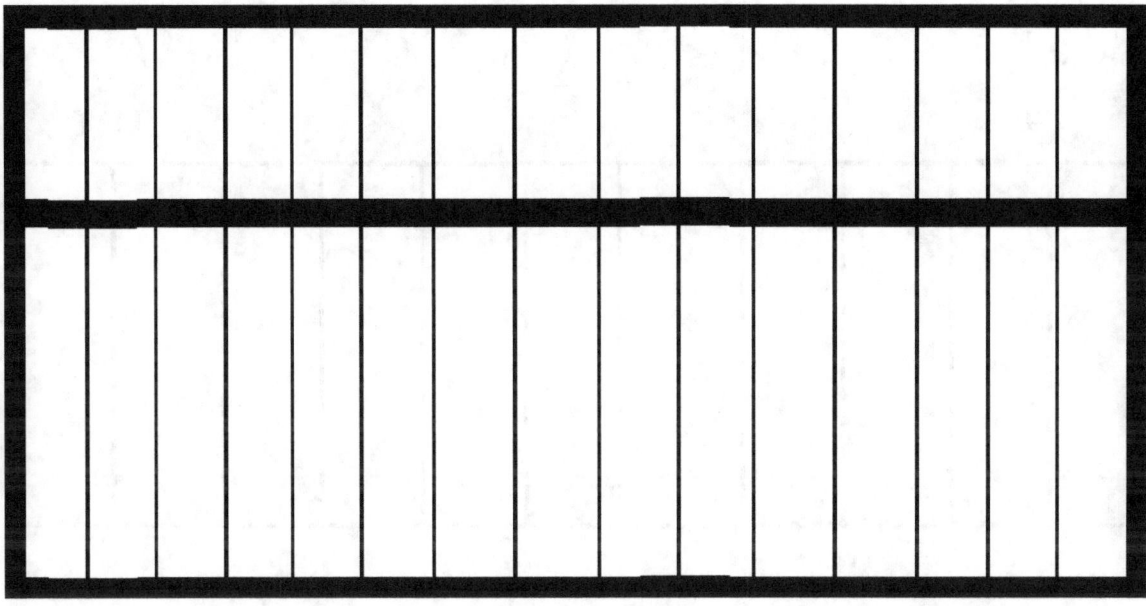

j. 3400+602-1009

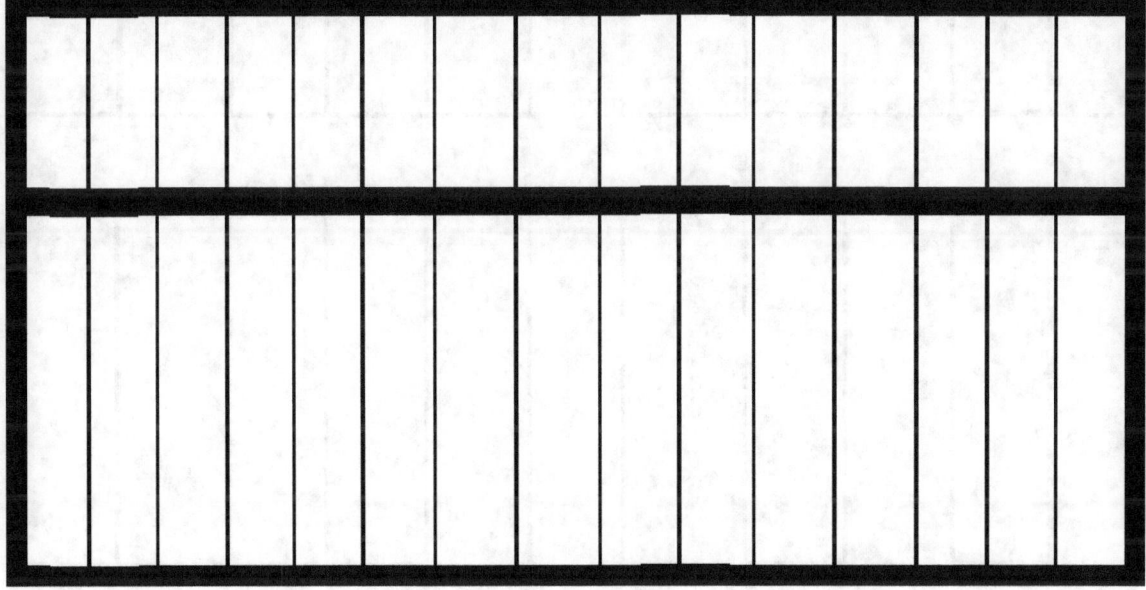

Test D, Show solution beads in final position.

Add a bar.

a. 16 x 7

b. 20 x 9

c. 13 x 4

d. *(65 - 33) x 19*

e. *12/4 x 64/8*

f. *219 / 3*

g. 33 x 47 +9 -18 / 7   *convert remainders to 2 decimal places*

h. (65 X 88) / 43

i. 23+19+37.9 + 3.5 - 12

j. $ .68 + $26.50 + $208 - $45.01

Complete all tests and then check the answers in this book, Subtract 2.5% for each wrong answer. An answer is wrong if any part of the answer is wrong. The answer may be placed anywhere on the Abacus, because you determine where the decimal point would be. There is no credit for completing more than what is asked for but you will give yourself an honorable mention! Add back 2% for each wrong answer you are able to correct. If you score less than 75%, you need to start this book and all the work again.

100%   Abacus Master

90%   Abacus Expert

80%   Successful Completion

Your Certificate is at the back of this book.

Let me know if you enjoyed learning how to use the Chinese Abacus.

Yours in Learning and Teaching,

Joe Salazar

Jws345@hotmail.com

Test A, Answers

a. 2 2 + 3 4 =   5 6

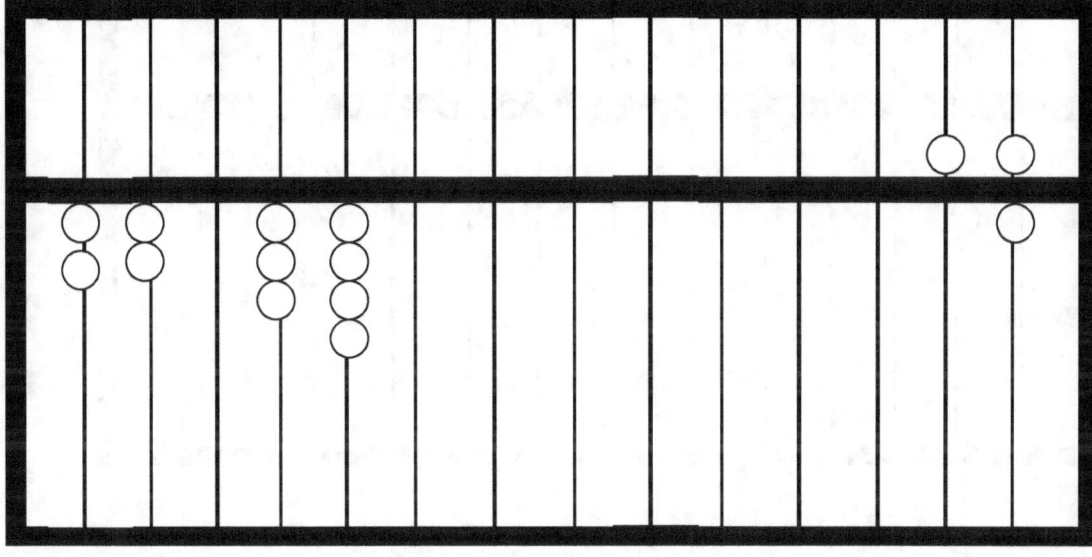

b. 1 6 + 2 6 =   4 2

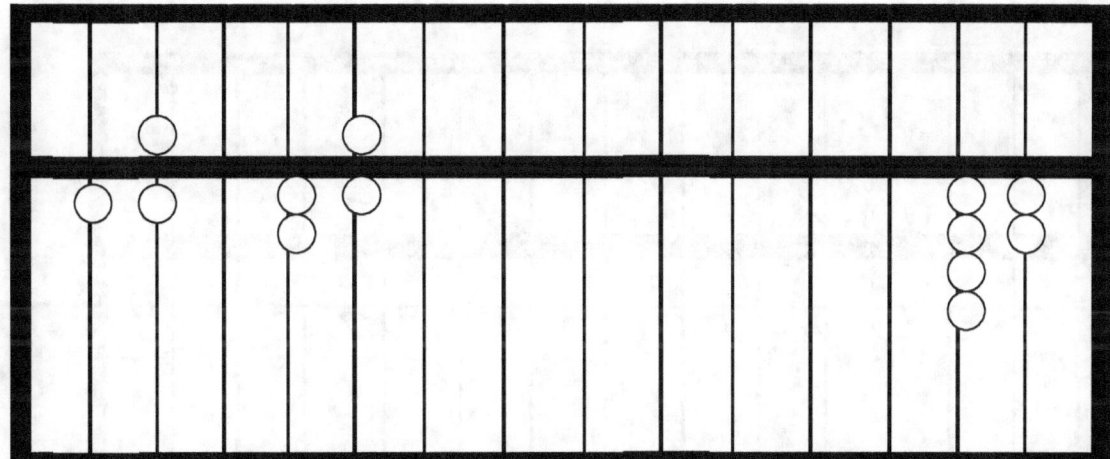

c. 104+220 =   3   2   4

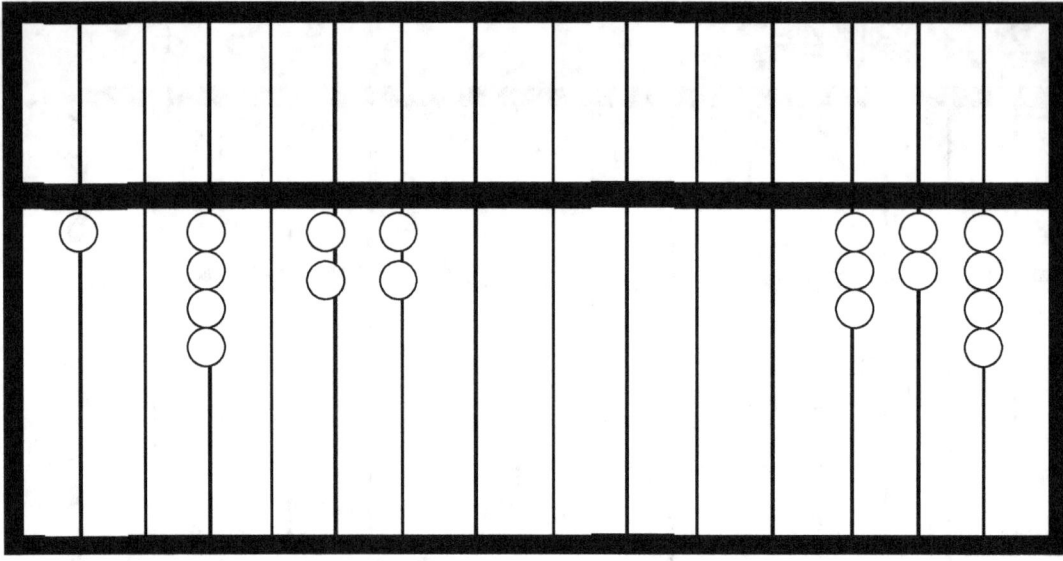

d. 30+70 = 100

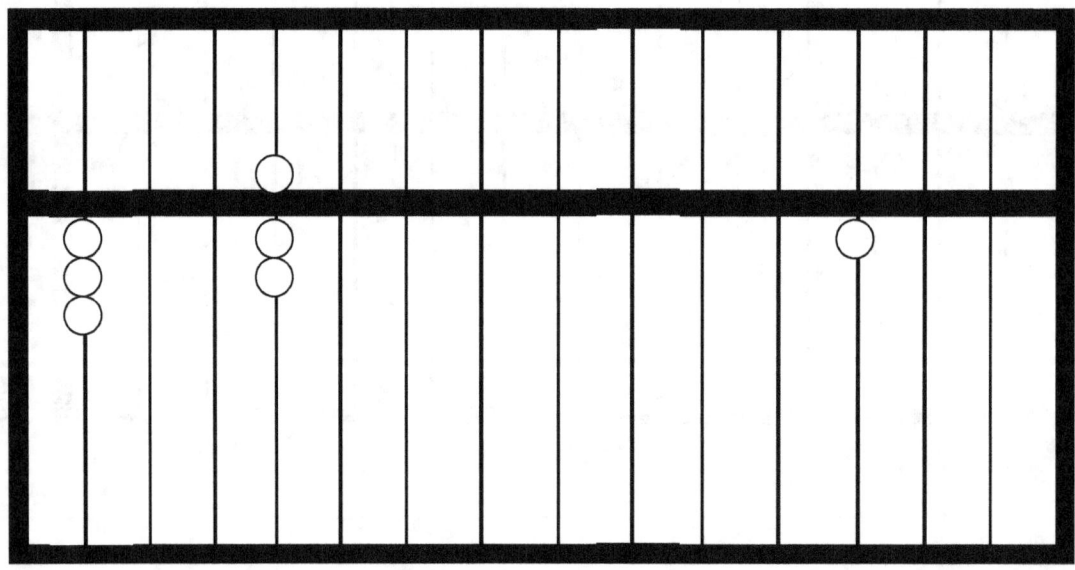

e. 4004+2030 = 6 0 3 4

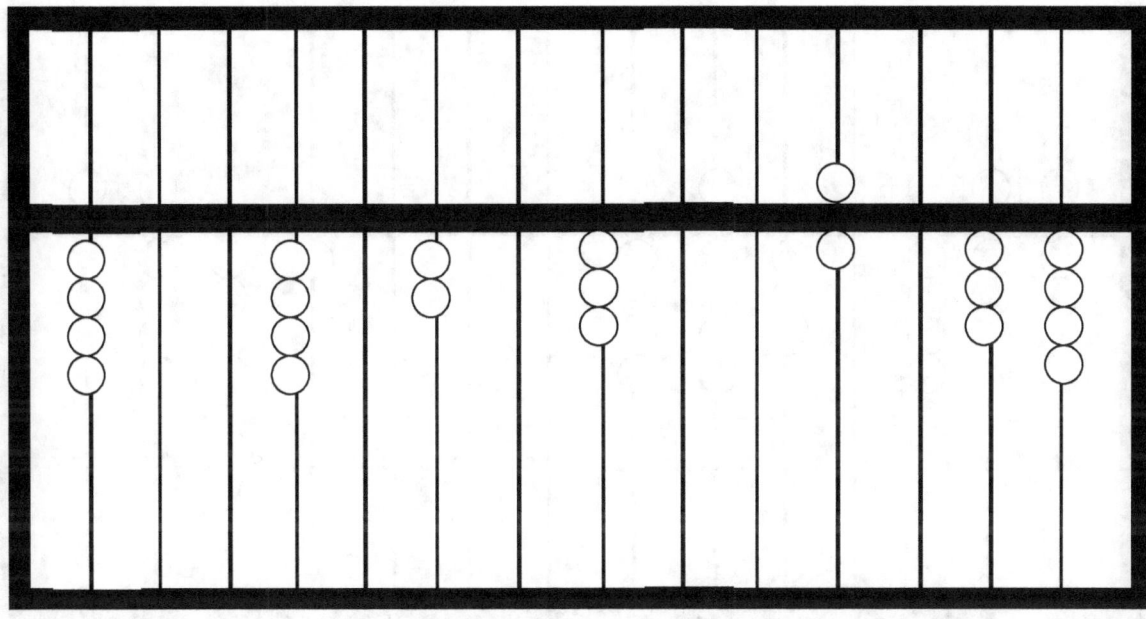

f. 180 + 810 = 990

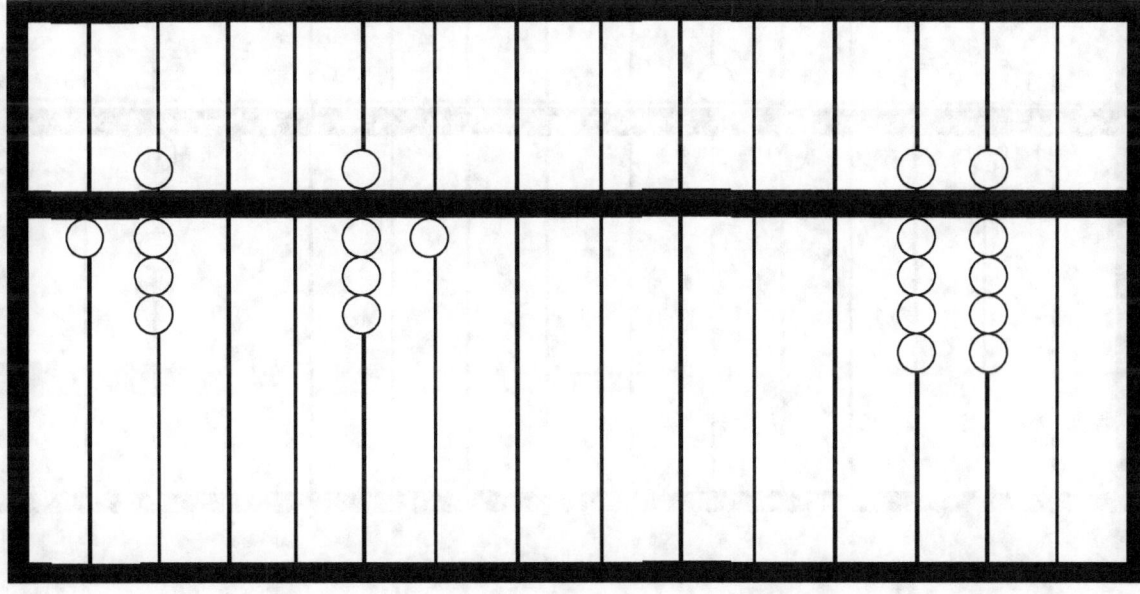

g. 999+499 = 1,498

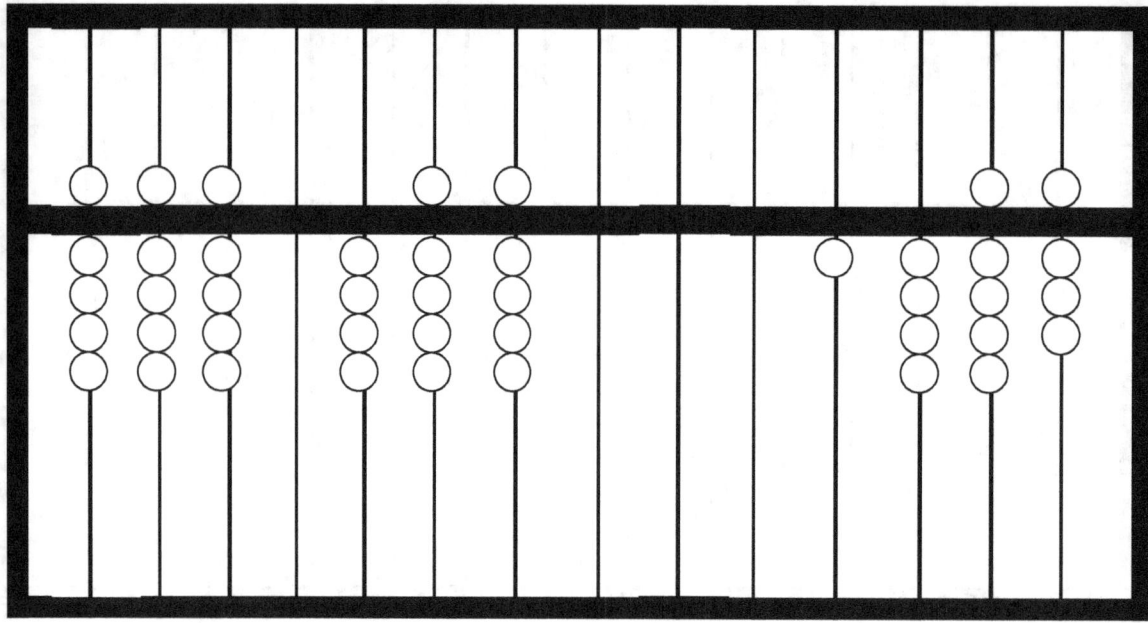

h. 115+15 = 130

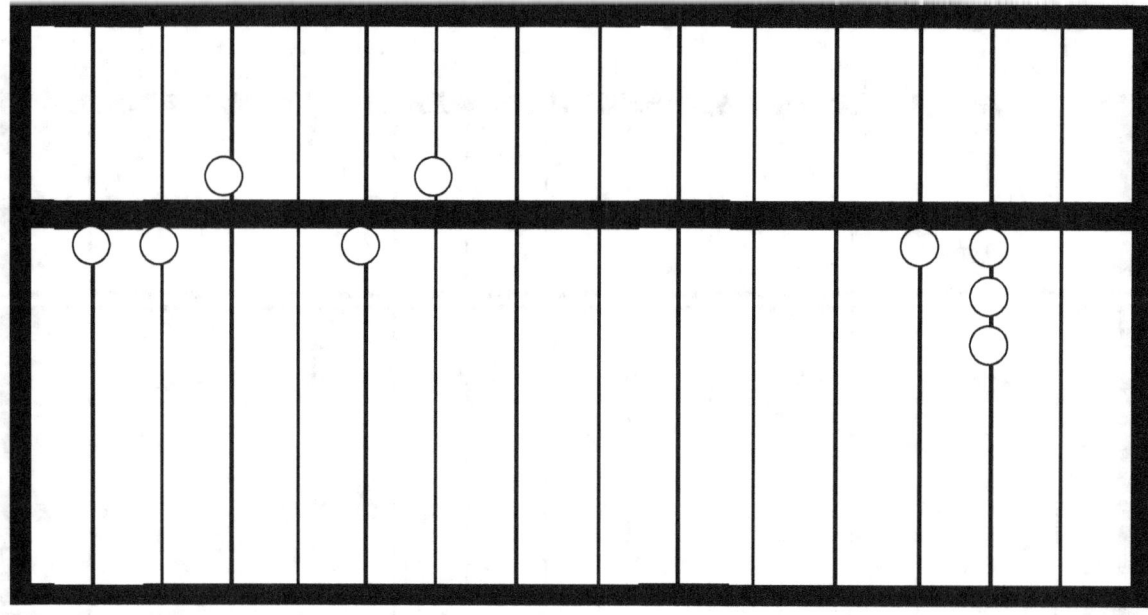

i. 203 + 219 = 422

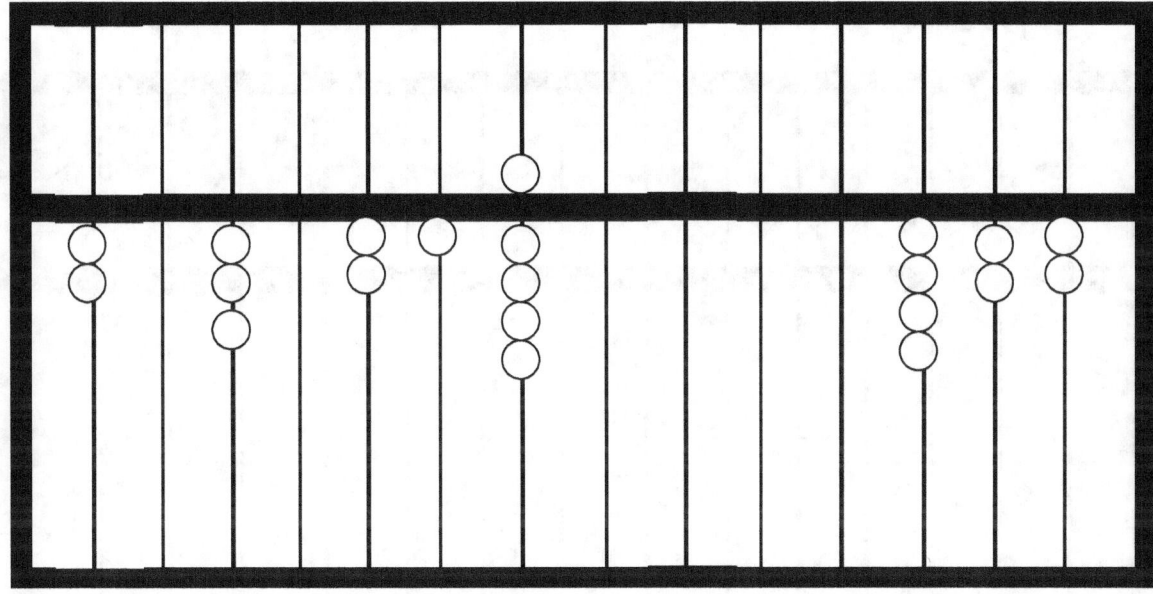

j. 4009 + 9090 = 13, 099

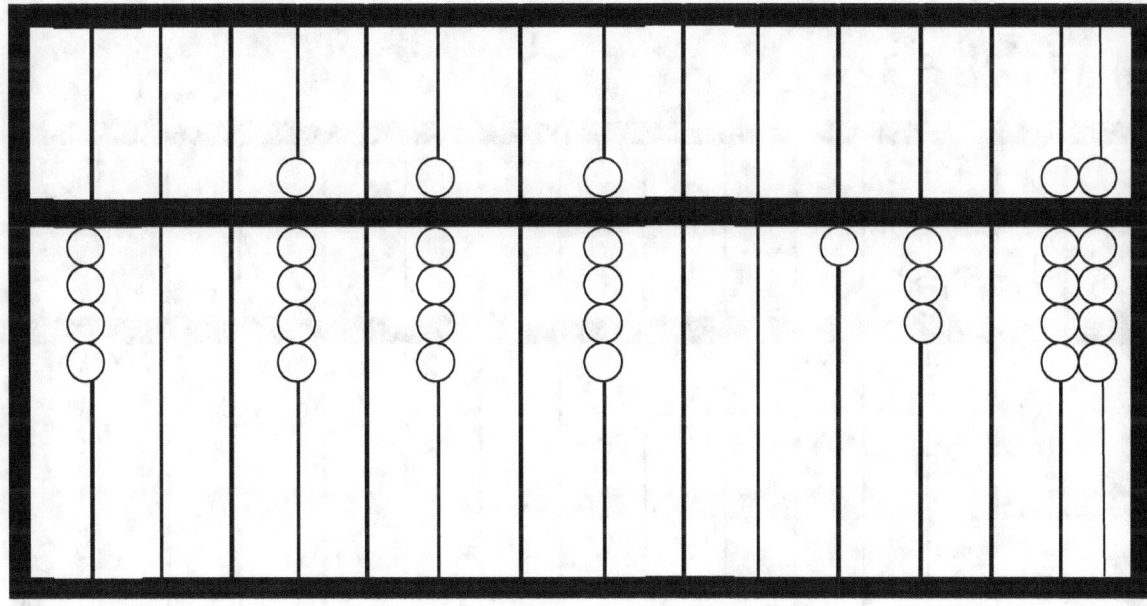

Test B, Answers

a. 35-18 = 17

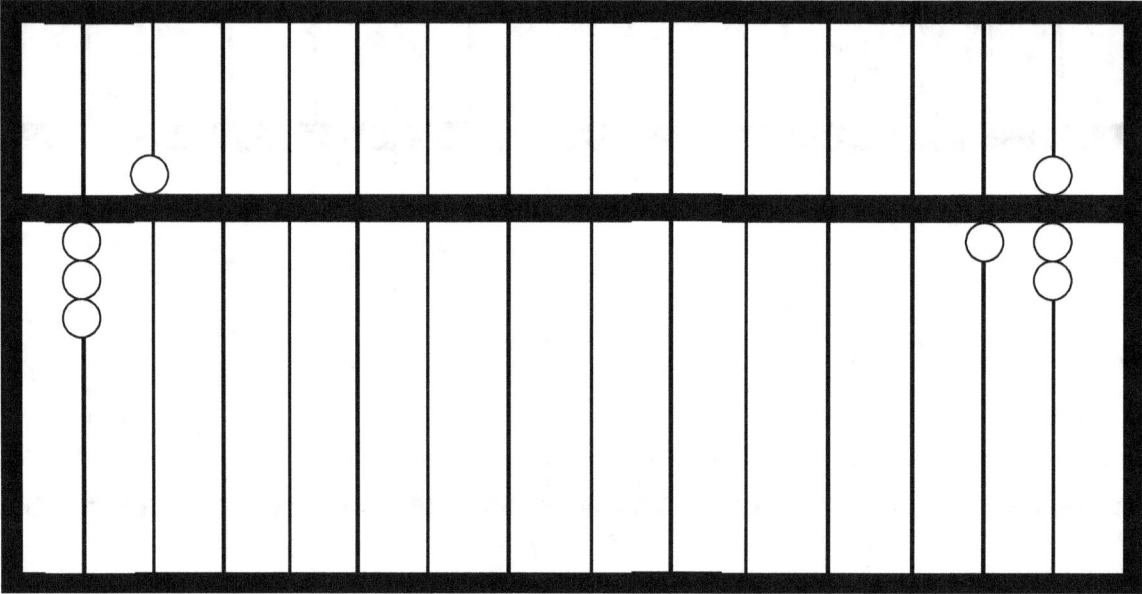

b. 355-28 = 327

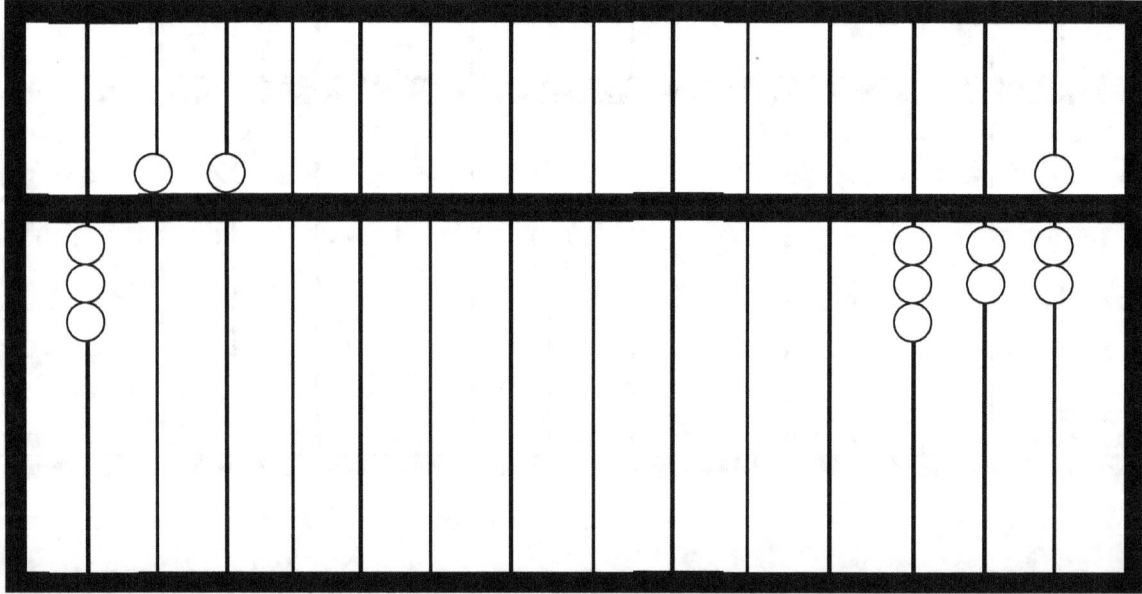

c. 298-120 = 178

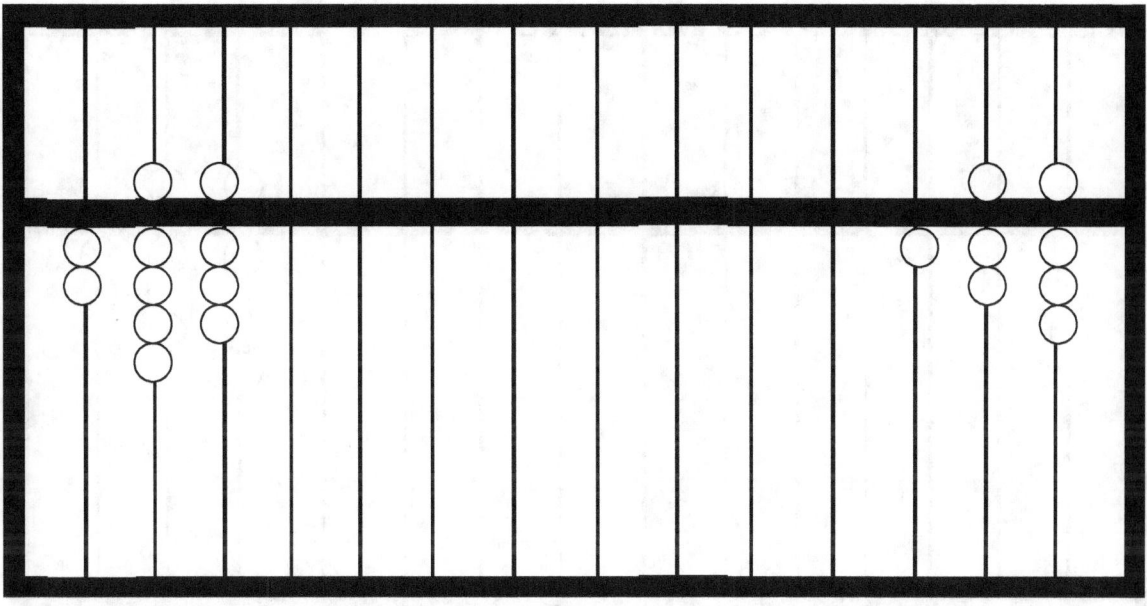

d. 507-203 = 304

e. 3001-2003= 998

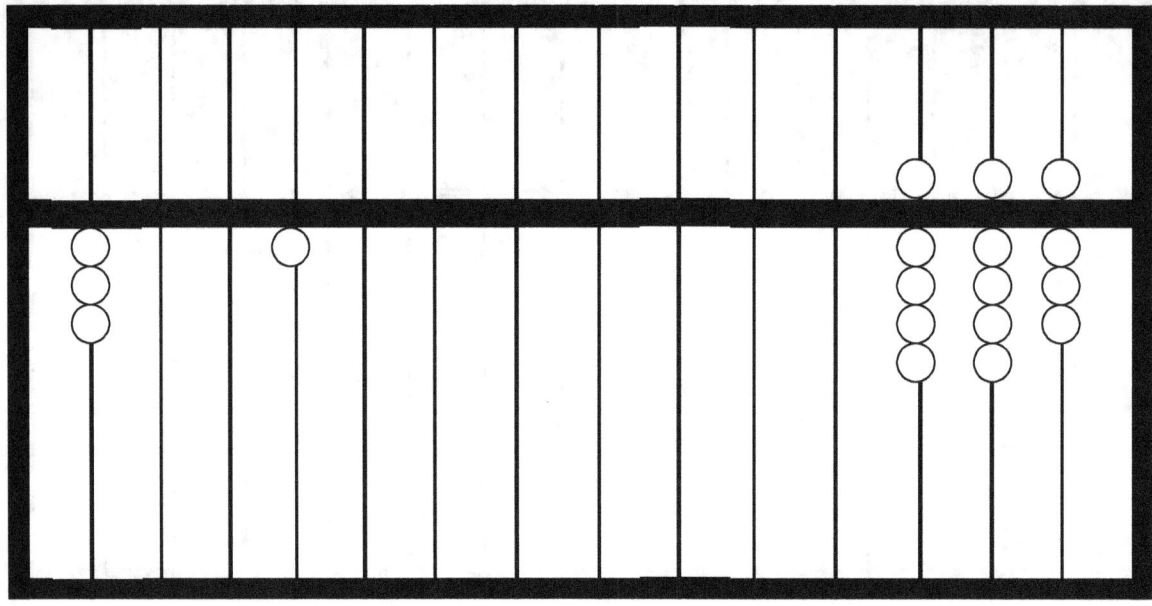

f. 100-97= 3

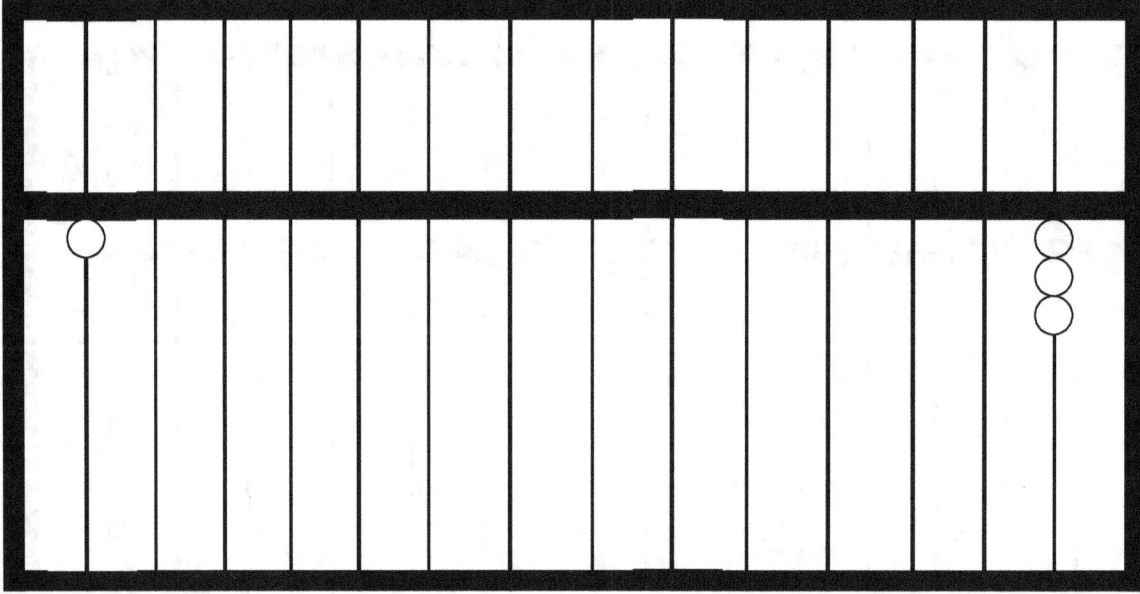

g. 222-199 = 23

h. 606-284 = 322

i. 903-605 = 298

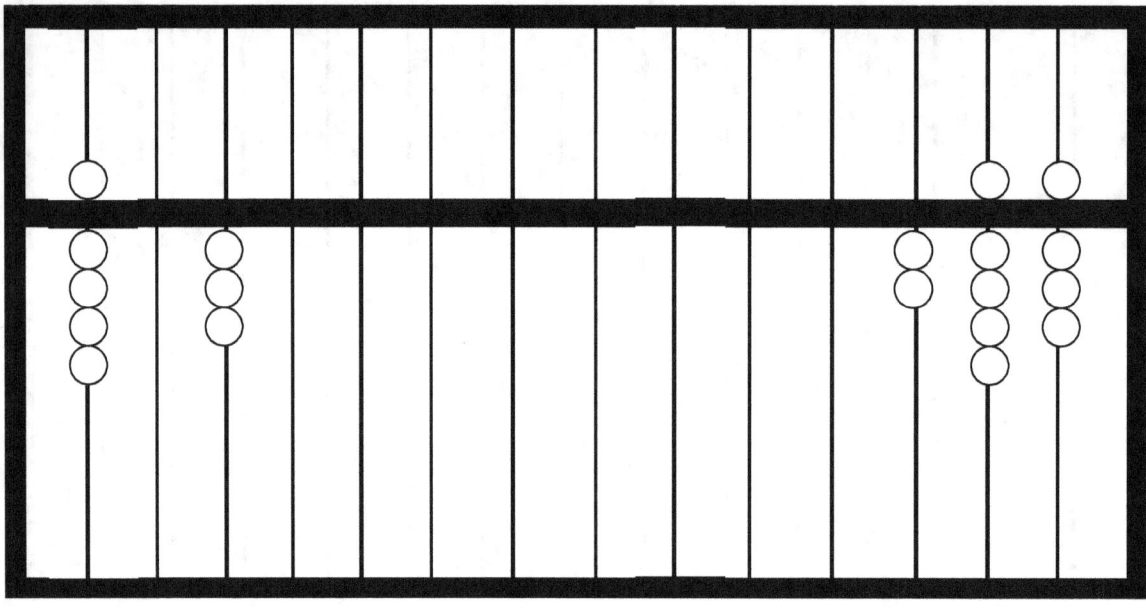

j. 505-209 = 296

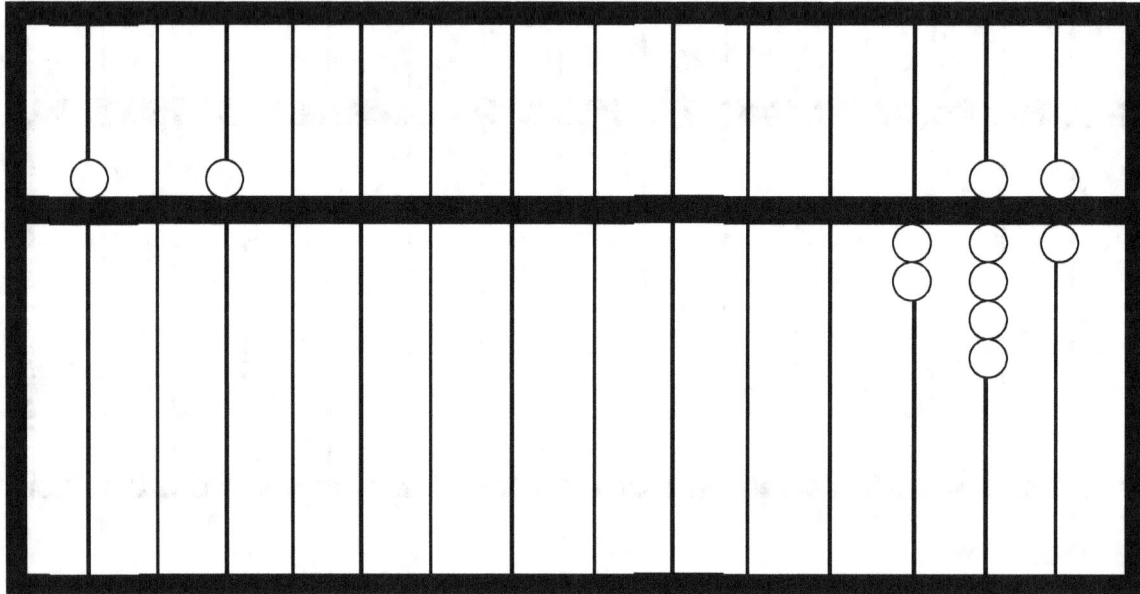

Test C, Diagram only the solution calculated on abacus

a. 23+18-9+17 =49

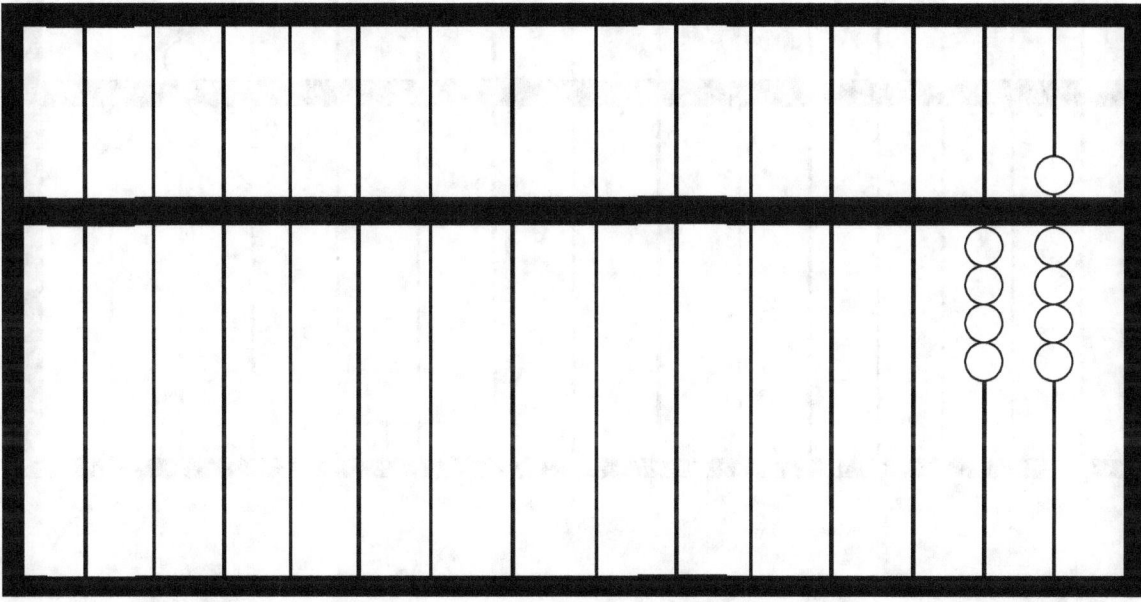

b. 304-210+190-12 = 272

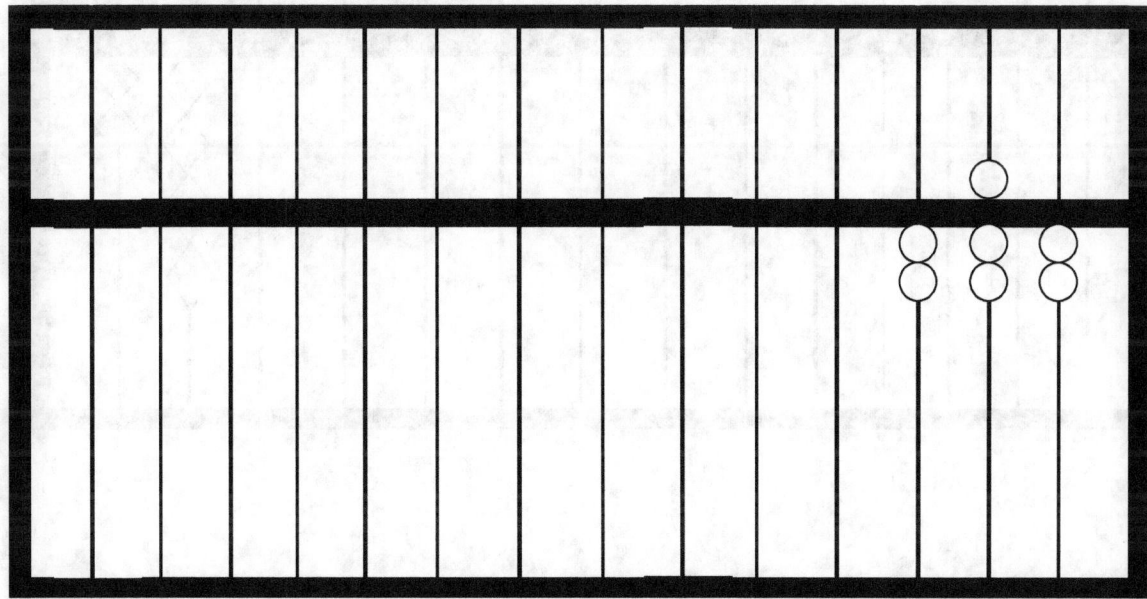

c. 213-31+180 = 362

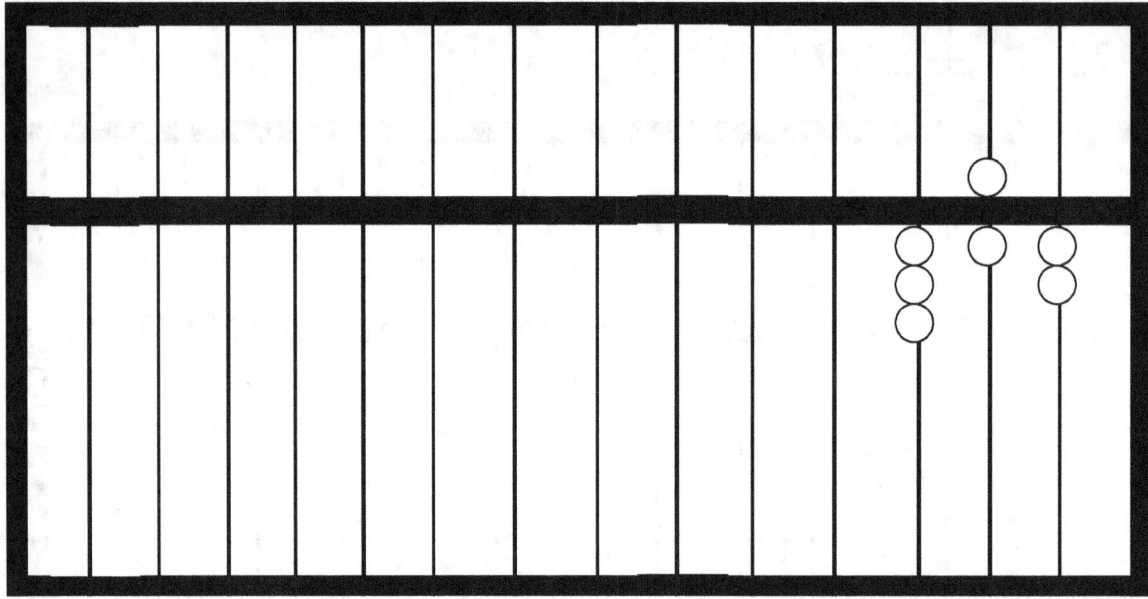

d. 108-9+19-26 = 92

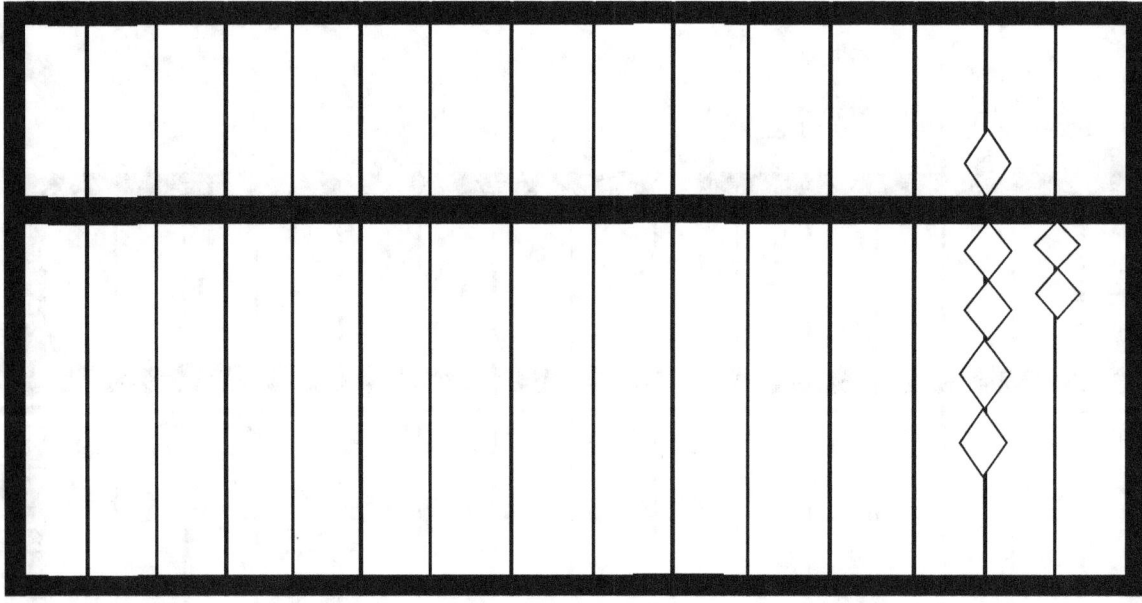

e. 999-18+54-22+75 = 1088

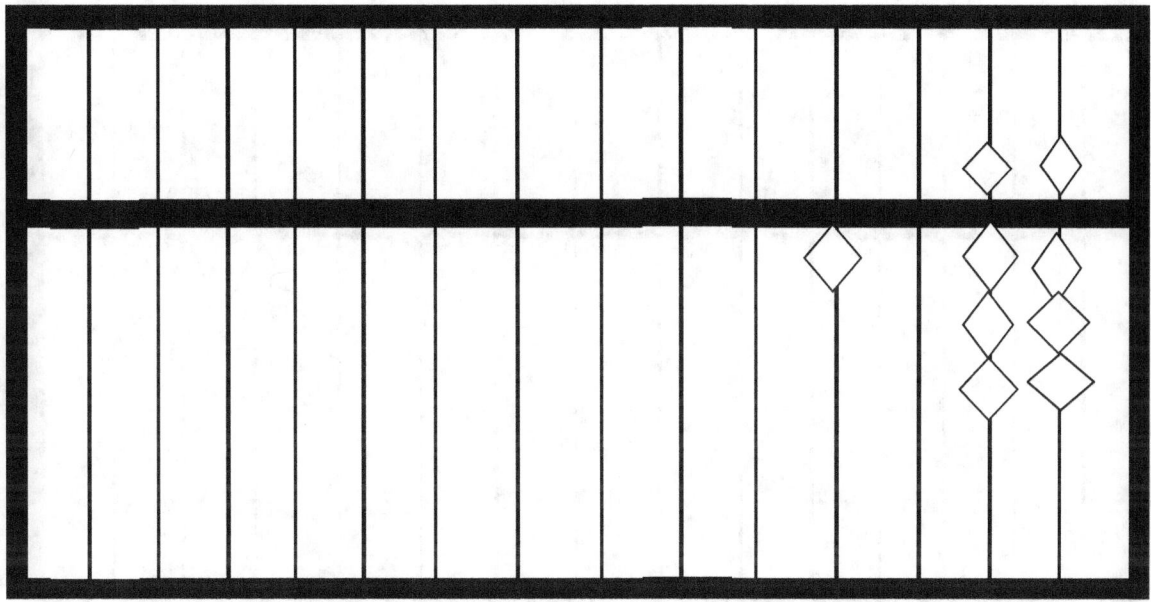

f. 211+112-121+212-2 = 412

g. 309-27+91-100 = 273

h. 209-87+91-17+20 = 216

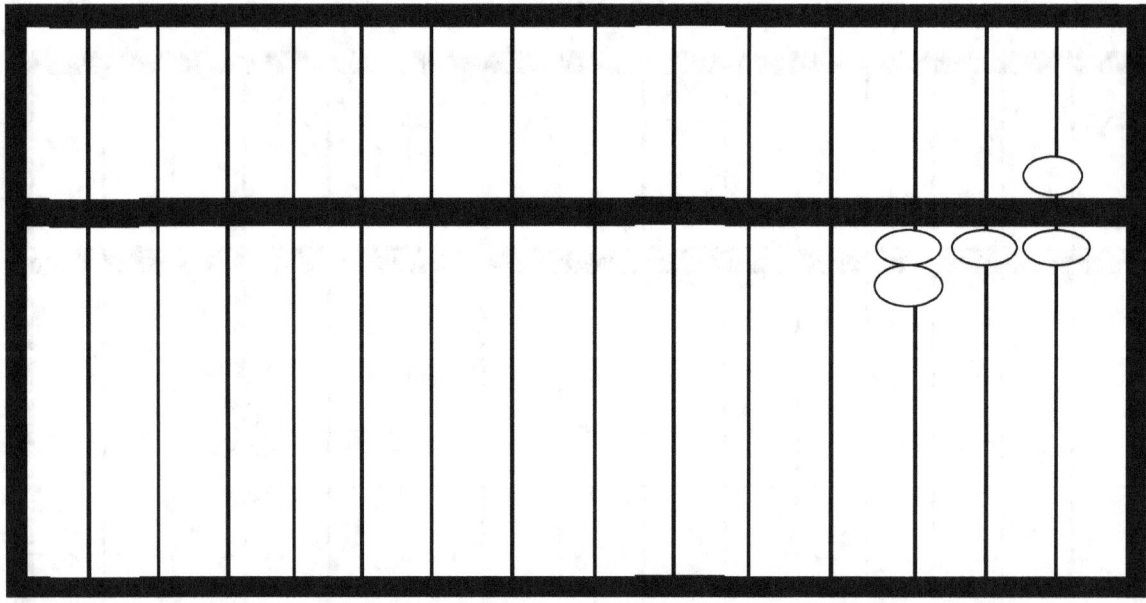

i. 120-19+109-15-22+390 = 563

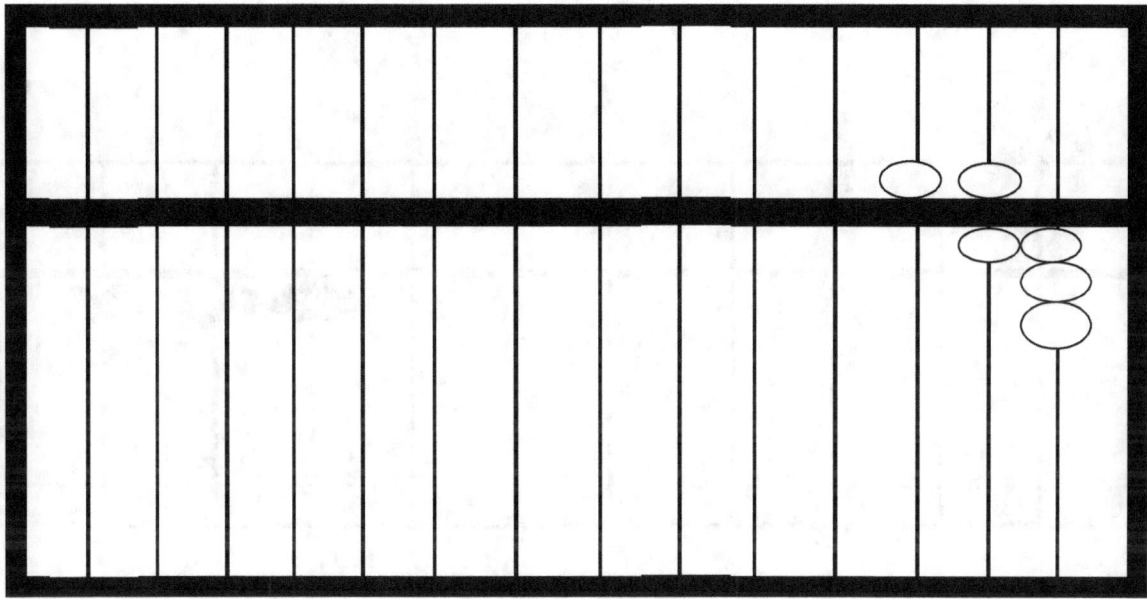

j. 3400+602-1009 = 2993

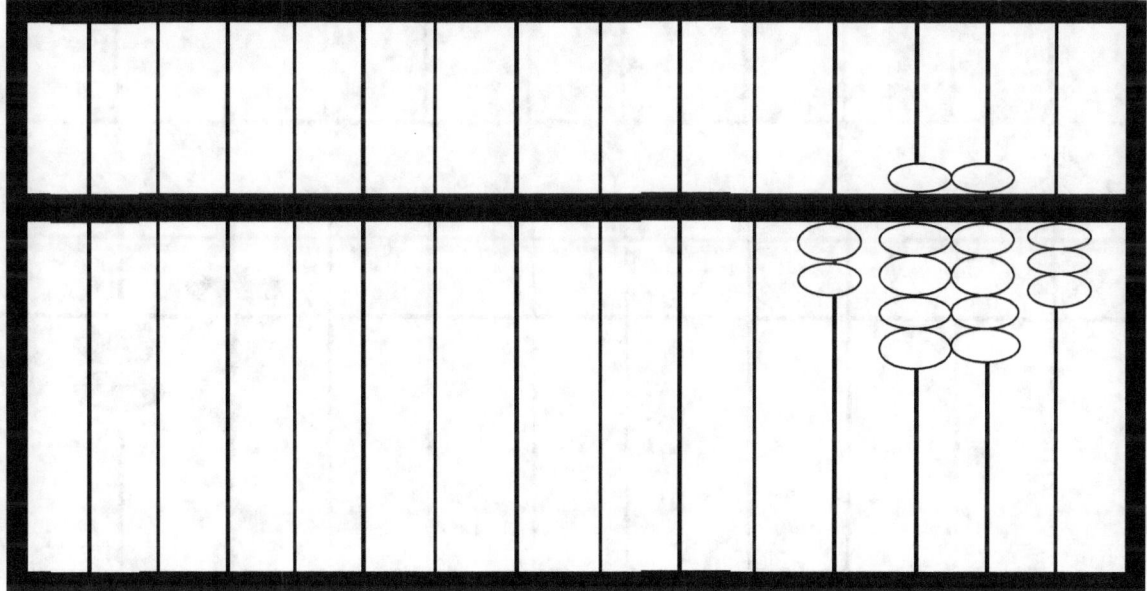

Test D, Answers

a. 16 x 7

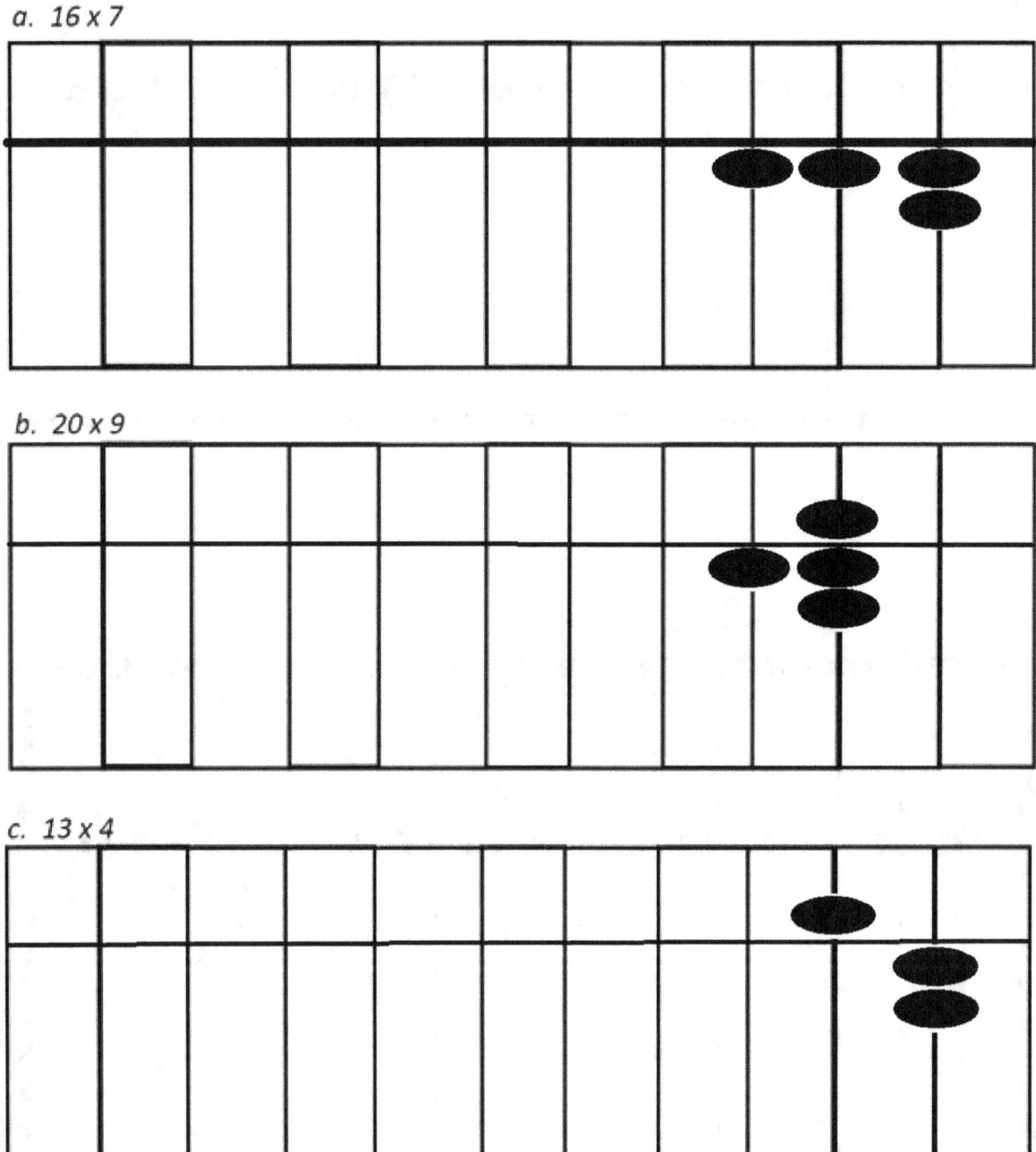

b. 20 x 9

c. 13 x 4

d. (65 - 33) x 19

e. 12/4 x 64/8

f. 219 / 3

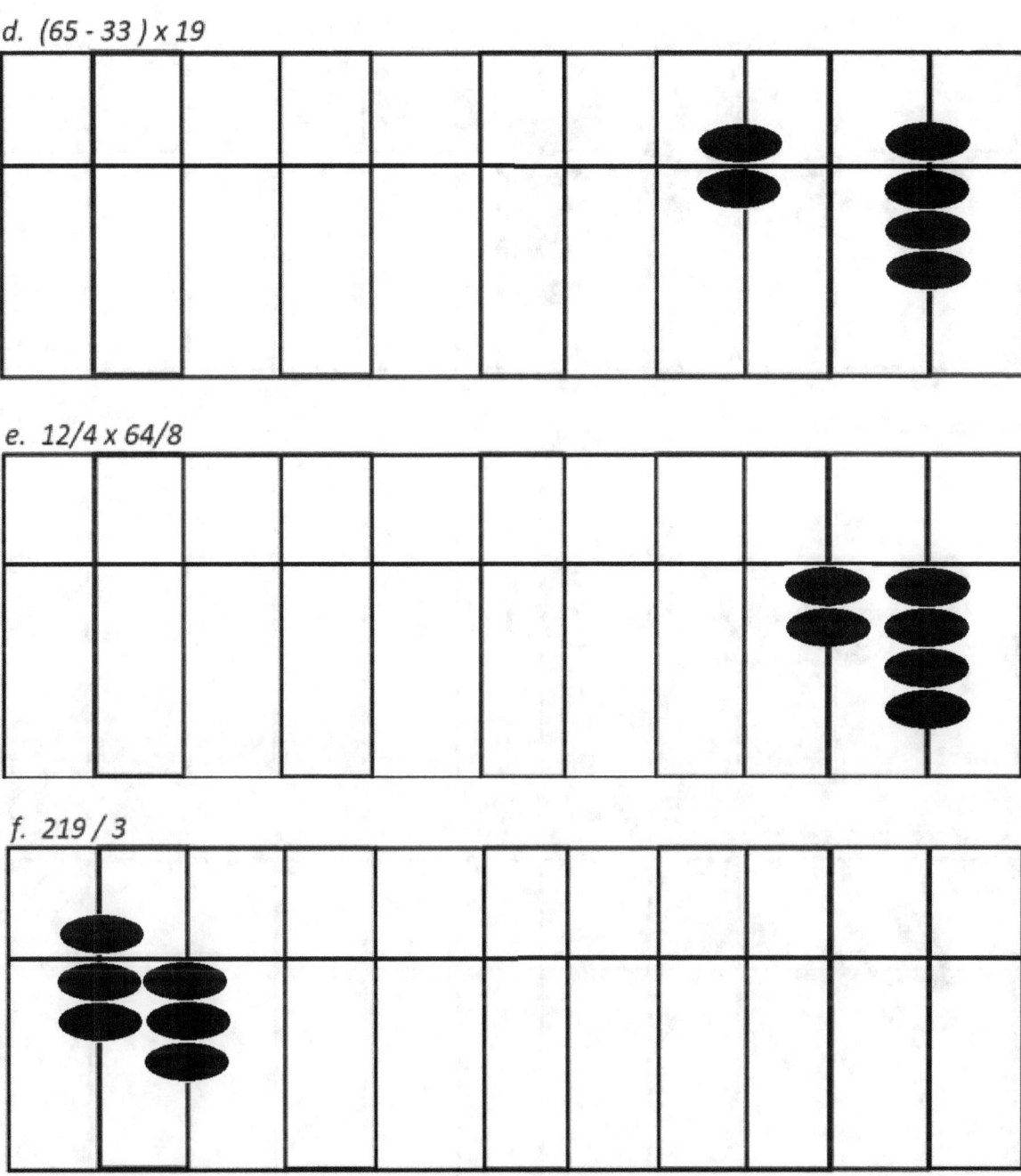

It is appropriate to plan a quotient from left to right because we don't know how many spaces we will need.

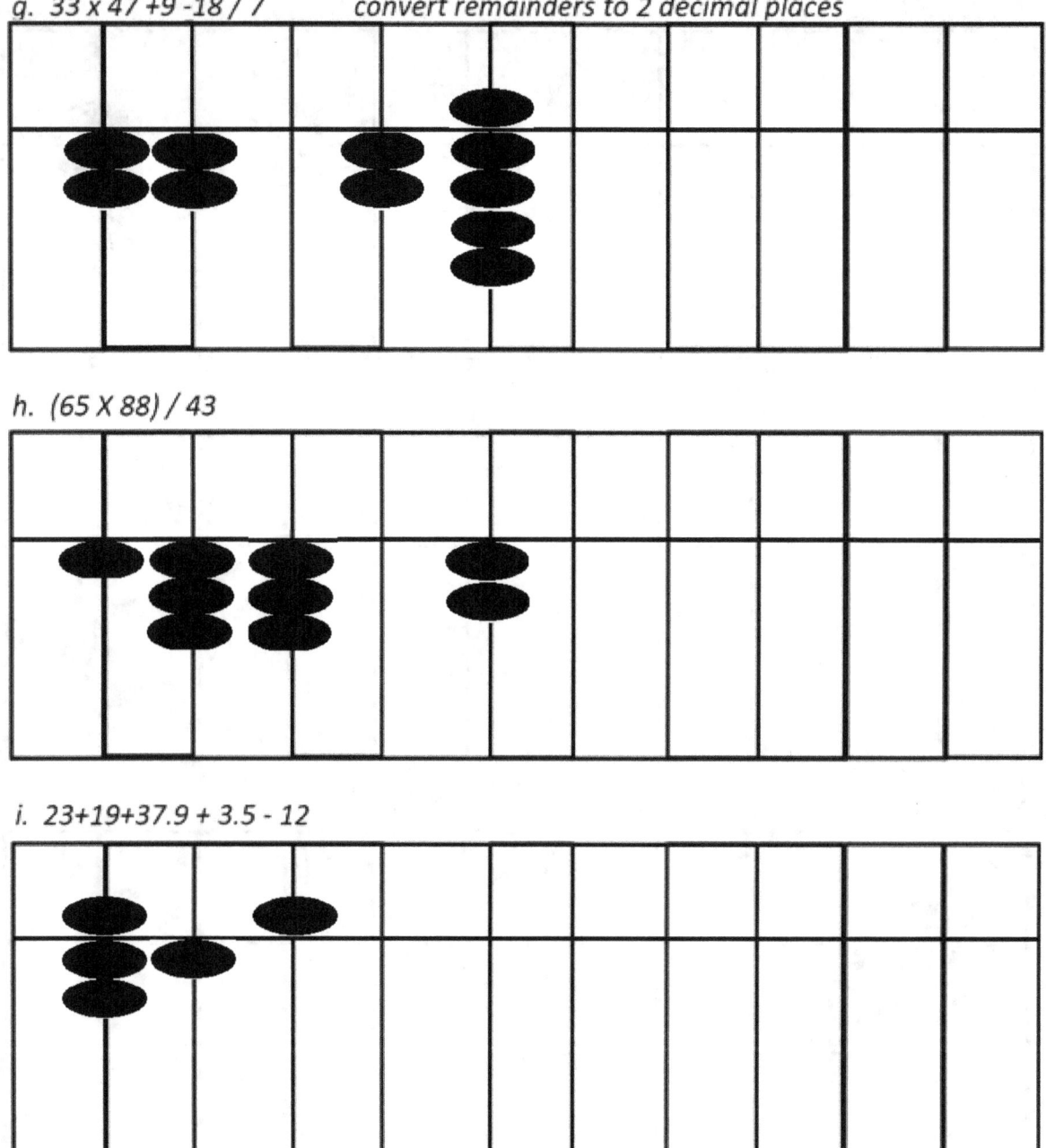

g. *33 x 47 +9 -18 / 7*   *convert remainders to 2 decimal places*

h. *(65 X 88) / 43*

i. *23+19+37.9 + 3.5 - 12*

Congratulations you have completed this course of, Chinese Abacus, Training.

The Author

Joseph W. Salazar

Born Dec. 9, 1940 Chicago Illinois

Enlisted in the U.S. Army at age seventeen, April 1958

Assigned to the Medical Corps and served six years, four of which were in Germany. I worked in many types of Army Medical Clinics and hospitals as a Medical Attendant and Technician. After leaving the Army I lived in Salt Lake City for two years before moving to California and worked as a Vocational Nurse, Driving Instructor, Martial Arts Teacher, and Traffic Violator School Instructor, while obtaining my B. S. in Health Science at San Jose State University.

I got my J. D. from Saratoga University Law School, Distance Learning Program.

My hobbies are Chess, Martial Arts, Archery and Flying Planes.

I taught the Accounting Clerk Program for Goodwill Industries for three years and operated my own Tax & Accounting business for forty years. I continue to operate my own business as a Public Accountant.

Joe Salazar

Prunedale California, August 2023

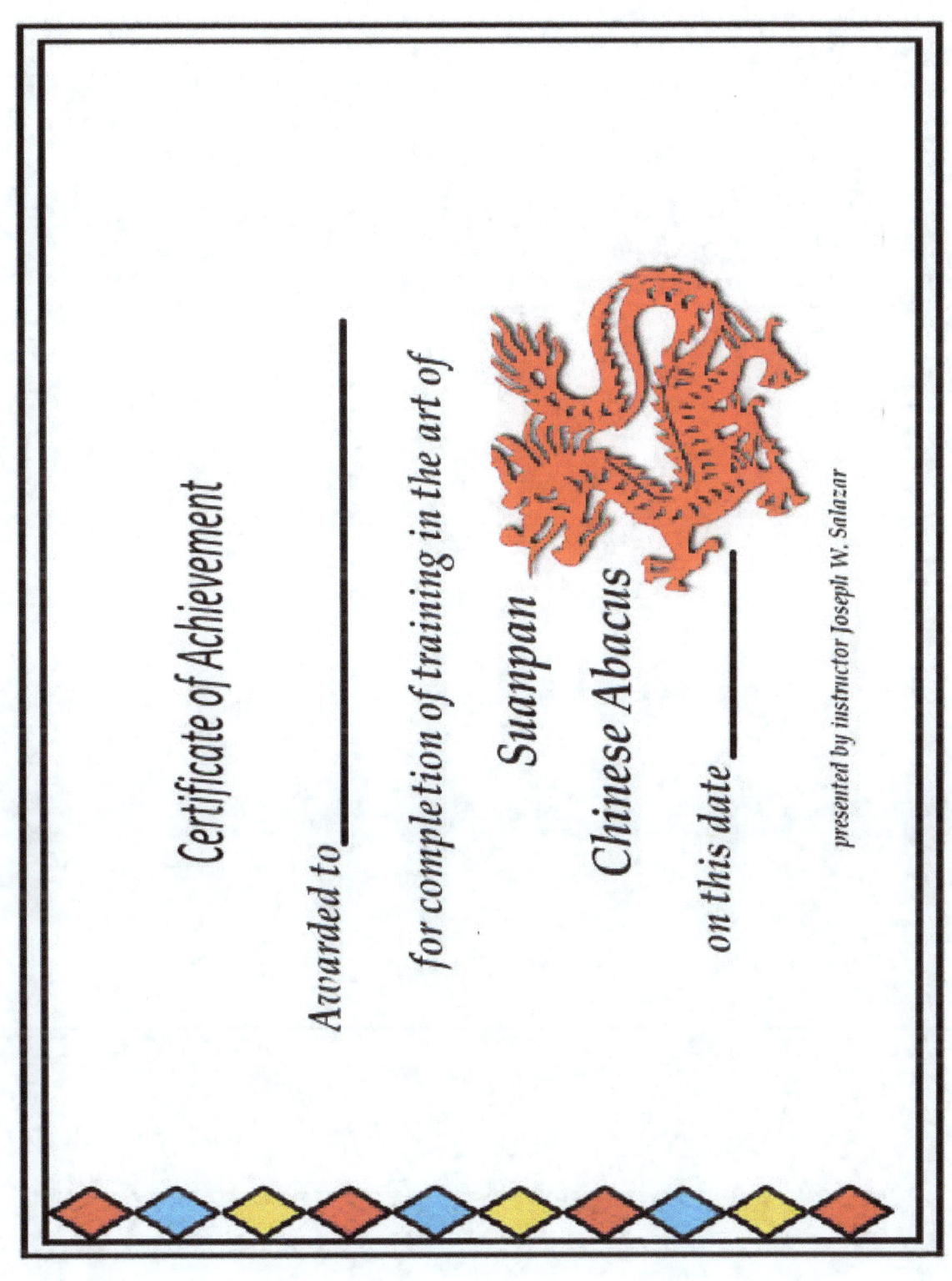

Certificate of Achievement

Awarded to _____

for completion of training in the art of

Suanpan

Chinese Abacus

on this date _____

presented by instructor Joseph W. Salazar